Site health and safety is not a nice to have option, it is a right for all engaged in construction. It is about education, empowerment and leadership, not process or policing. Without all involved sharing the belief we should and can have safe sites, hazardous behaviours will emerge that history tells us results in needless deaths and injuries.

This easy-to-use reference guide gives an overview for all who may be working on-site, to prepare you for working safely, and provides advice on what to do when you come across commonly encountered health hazards.

The information will also assist designers to appreciate the site hazards which they need (under CDM2007) to identify, eliminate where possible and then reduce the level of residual risk.

Everyone on site should find it helpful and useful. It also takes account of recent changes in health and safety legislation, for example The Control of Vibration at Work Regulations 2005, and emphasises the need for all those involved in construction to identify, assess and manage risks.

I am pleased that CIRIA has produced this revised handbook for the benefit of all those in the construction industry. I commend it to your use.

Keith Clarke
Chairman of the Health and Safety Committee
of the Construction Industry Council

Site health handbook (fourth edition)

Pendlebury, M, Brace, C, Gibb, A, Gyi, D and Gilbertson, A L

CIRIA

C670   © CIRIA 2008   ISBN 978-0-86017-670-1

First published 2004 (C629)

British Library Cataloguing in Publication Data

A catalogue record is available for this book from the British Library

Published by CIRIA, Classic House, 174-180 Old Street, London EC1V 9BP, UK

Keywords		
Health and safety, site management, regulation, construction management, project management, respect for people		
Reader interest	**Classification**	
Health and safety, construction occupational health, site management, CDM2007	AVAILABILITY	Unrestricted
	CONTENT	Advice/guidance
	STATUS	Committee-guided
	USER	Construction professionals, architects, engineers, designers, surveyors, planners, site managers, site supervisors, site workers, construction managers, contractors, project managers/directors, local authority staff, CDM co-ordinators, regulators, construction clients

This publication is designed to provide accurate and authoritative information on the subject matter covered. It is sold and/or distributed with the understanding that neither the authors nor the publisher is thereby engaged in rendering a specific legal or any other professional service. While every effort has been made to ensure the accuracy and completeness of the publication, no warranty or fitness is provided or implied, and the authors and publisher shall have neither liability nor responsibility to any person or entity with respect to any loss or damage arising from its use.

All rights reserved. No part of this publication may be reproduced or transmitted in any form or by any means, including photocopying and recording, without the written permission of the copyright holder, application for which should be addressed to the publisher. Such written permission must also be obtained before any part of this publication is stored in a retrieval system of any nature.

If you would like to reproduce any of the figures, text or technical information from this or any other CIRIA publication for use in other documents or publications, please contact the Publishing Department for more details on copyright terms and charges at: publishing@ciria.org Tel: 020 7549 3300.

CIRIA C670

UNIVERSITY OF STRATHCLYDE

30125 00808442 7

London, 2008

ML

UNIVERSITY OF STRATHCLYDE LIBRARIES

Books are to be returned

Site health handbook

Second edition

Prepared by

Martyn Pendlebury

Charlotte Brace

Alistair Gibb

Diane Gyi

Revised by

Alan Gilbertson

CIRIA *sharing knowledge ▪ building best practice*

Classic House, 174–180 Old Street, London EC1V 9BP
TELEPHONE 020 7549 3300   FAX 020 7253 0523
EMAIL enquiries@ciria.org   WEBSITE www.ciria.org

UNIVERSITY OF STRATHCLYDE

– 8 AUG 2008

UNIVERSITY LIBRARY

D
690·22
SIT

2   C670 Site health handbook

Acknowledgements

This guide was produced as a result of CIRIA Research Project 673, *Handbook to site health – guidance for site workers* by Martyn Pendlebury, Charlotte Brace, Alistair Gibb and Diane Gyi of Loughborough University.

This revision of CIRIA's *Site health handbook* was carried out by Alan Gilbertson. Work has been funded by CIRIA who wishes to express its thanks to all who contributed to this and the previous edition.

The original project was funded by the Department of Trade and Industry, through its Partners In Innovation (PII) programme, CIRIA's Core members, Rail Link Engineering and the Construction Health and Safety Group. The project steering group comprised:

Graeme Cox (chair)	Rail Link Engineering
Annabel Bartlett	Mott MacDonald Ltd
Jason Bingham	Davis Langdon Consultancy
Bob Blackman	Transport & General Workers Union
Richard Boland	Health and Safety Executive
Duncan Calder	Zurich Risk Services
John Davey	Arup (representing the Construction Health & Safety Group)
Anthony Edwards	Balfour Beatty plc
Denis Hands	Construction Industry Training Board
Tony Hart	British Nuclear Fuels plc
Robert King	Engineers Employers Federation
Tony Metcalfe	Wates Group Limited
Andrew Sneddon	Construction Confederation
Peter Twinam	Arup (representing CHSG)
Tony Wheel	Carillion plc

CIRIA and the authors also thank Dr Andrew Colvin, Institute of Occupational Medicine, Edinburgh, and Nick Charlton-Smith, Association for Project Safety (formerly Association of Planning Supervisors), for their medical advice in Section 3.

CIRIA is particularly grateful to David Lambert of Kier Group, David Watson of WSP and Graham Leech of Balfour Beatty Management for assistance with the 2008 update.

Contents

Contents

Summary

This *Site health handbook* provides practical advice for operatives, supervisors and managers working in construction as well as designers who are making decisions about construction.

The guide identifies how to recognise risk and minimise the impact of construction sites and operations on your health and that of your colleagues. The guide is intended to be a reference and a training aid.

The guide has four sections:

- **Section 1:**   Introduction to site health
- **Section 2:**   Site health: the basics
- **Section 3:**   Health issues at work
- **Section 4:**   Further help and information.

The *Site health handbook* is presented as a companion guide to C669 *Site safety handbook* (CIRIA, 2008).

Summary

When planning the control of health risks, the decision process for identifying the most appropriate control should be fairly formal. The list below gives the accepted hierarchy of control for protecting health and safety:

Eliminate – remove the substance or activity that is high risk (or substitute it with a less hazardous substance or activity)

Reduce – change the degree of risk from the substance or the activity

Inform – develop safe working procedures and provide effective training for handling the level of risk

Control – keep levels of the hazardous substance or activity as low as possible and maintain these levels. Provide personal protective equipment (PPE) only as a last resort.

The appropriate control is usually a combination of engineering, technical, procedural and behavioural controls.

 Remember: PPE is **not** the first line of defence for good health.

The target audience

Who should read this guide?

This guide provides practical advice for all people engaged or working in construction, including those with a responsibility or interest in the management of their workers' health.

The task of preventing or minimising the risks to health at site level demands high standards of health awareness and education. This guide provides practical information about health hazards and identifies how individuals can avoid or minimise risks to their health.

It is primarily aimed at:

- workers
- supervisors
- line managers.

Additionally, there is also a need to improve the health culture within the industry. To help achieve this objective, this guide should be brought to the attention of construction professionals including:

- construction planners within contractors' main offices
- contractors and principal contractors
- project managers/directors
- designers (including quantity surveyors, specifiers and buyers)
- local authority staff
- construction managers
- CDM co-ordinators
- regulators
- construction clients.

This guide covers health issues for construction workers and others who may be affected by construction work. The hazards that they may come into contact with during construction work are explained.

Statistics and pictures are used to show the short- and long-term effects to health and how ill-health can impact on individuals and their families.

The **principal objectives** of the guide are to:

- **set out** why it is important to understand the hazards and how they can be avoided or the level of risk reduced

- **educate and inform** workers, their supervisors and managers about the health risks associated with common construction activities

- **provide guidance** that can help reduce or avoid the risks to workers' health associated with hazardous materials and conditions

- **summarise** the basic health requirements at a site level so that workers and their supervisors are aware of their duty to ensure that a healthy working environment is maintained throughout the construction process

- **advise** that construction projects, irrespective of location, size or nature, offer potential risks to the health of site workers, and that site inductions about workplace hazards – before starting work – are essential

- **raise awareness** of the issues that affect good health. The induction process also helps workers understand the impact their work can have on current and long-term health

- **improve** the long-term health prospects for construction workers and other people in the industry.

> **All construction professionals should discharge their duties under the Management Regulations and CDM2007 by working together to reduce health problems in construction by improved design, planning and site management.**

Coverage of this guide

Readers should be clear about the scope and limitations of this guide and in particular that:

- it identifies risk and gives guidance, but detailed advice may need to be found from other sources

- the guide should not replace meetings and discussions with regulatory and medical authorities and other key interested parties

- the guide only summarises legislation. For more detail you should contact your company's medical or legal advisors and regulators such as the HSE

- in all instances when dealing with the issues covered, do not take action beyond your knowledge and ability. If in doubt, seek specialist advice.

This book is intended to be used as a reference and a training manual.

The guide is split into four main sections:

Section 1 introduces the benefits of good health practice and why it should always be adopted. It examines the most common construction processes and identifies the materials that can affect your health.

Section 2 gives you some straightforward tips on what you should do before you start on site and what you should do as soon as you start a new job. This section also discusses the basics of how to identify hazards and how to take action to eliminate or avoid them, or reduce the risks arising from them.

Section 3 describes the health hazards that can affect your body when working on a construction site, eg manual handling, vibration, exposure. Descriptions are given explaining when you are at risk, why you are at risk and what you can do about it.

Section 4 is about construction activities involving hazardous materials, and includes an annotated bibliography. It also contains a list of useful organisations to contact if you have been affected by ill-health or need further information.

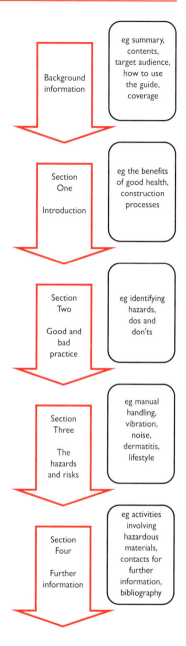

How to use this guide

The guide should be used as part of an integrated health management strategy that includes:

- setting targets for health improvement
- implementing risk assessments and control measures
- tool box talks
- informing, instructing and training
- applying effective surveillance
- improving understanding
- monitoring performance.

This guide is closely related to its sister document CIRIA C669 *Site safety handbook*.

Throughout the text, important information has been highlighted with the use of bullet points and text boxes. The book has also been illustrated with the use of photographs. However, please note that these are for illustration only and do not always show complete examples of best practice.

Throughout this guide, the symbols below have been used to help identify the types of information being provided and to simplify its use:

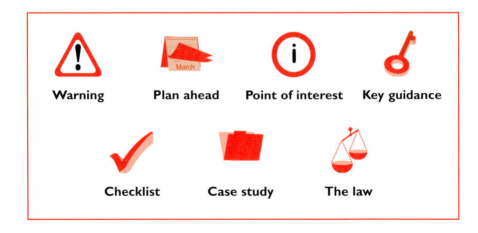

Warning   **Plan ahead**   **Point of interest**   **Key guidance**

Checklist   **Case study**   **The law**

Term	Definition
Anthrax	An acute infectious disease of farm animals caused by bacteria and which can be transmitted to humans by contact with animal hair, hides or excrement.
Asbestos	A group of naturally occurring minerals used in building materials etc.
Asbestosis	A chronic lung disease caused by exposure to asbestos.
Aspergillosis	A group of lung conditions caused by fungus.
Asthma	A narrowing of the airways making it difficult to breathe.
Carcinogen	A substance that causes cancer.
CE mark	The CE mark is given to products (such as PPE) that should be of a suitable standard so as to be safe for the people using them.
Cirrhosis	Liver disease usually caused by drinking excessive amounts of alcohol.
Cold stress	The term used to describe the lowering of your body temperature due to long periods of time in a cold environment.
Conjunctivitis	Inflammation of the eye, making it red and swollen, and producing pus. Caused by bacteria or viruses, allergy, or physical or chemical irritation.
Dermatitis	Inflammatory condition of the skin that is caused by outside agents. It is caused by contacts with irritants such as acids, alkalis, solvents and detergents.
Decompression illness	Complaint caused by breathing compressed air of between 0.25 bar and 3.5 bar above normal pressure. It can occur during air range diving (inland, offshore, inshore) mixed gas diving (mostly offshore) or when using compressed air in tunnelling.

Glossary

Term	Definition
Fibrosis	A health problem caused by breathing in silica dust for long periods. Fibrosis is a hardening of the lung tissue, making it hard to breathe.
HAV	Hand-arm vibration. vibration that reaches your hands when you are working with hand-held power tools or machinery.
HAVS	Hand-arm vibration syndrome. The injuries caused by HAV. These can include VWF (vibration white finger), damage to your nerves, muscles, bones and joints as well as your blood circulation.
Heat stress	The overheating of your body due to working in a hot environment for a long time.
Hepatitis	Inflammation of the liver caused by viruses, toxic substances, or immunological abnormalities.
Iris	The coloured part of your eye.
Jaundice	From skin-borne infections. Yellowing of the skin or whites of the eyes, indicating excess bilirubin (a bile pigment) in the blood.
Leptospirosis	Also known as Weil's disease. An infectious disease, caused by bacteria from the urine of rats, cattle, foxes, rodents and other wild animals – rats and cattles being the most common form of transmission in the UK. The disease begins with a fever and may affect the liver (causing jaundice) or brain (causing meningitis). It can also affect the kidneys.
Manual handling	Using your body strength to lift, lower, push, carry, pull, move, hold, or restrain equipment, products or tools.
Mesothelioma	A tumour, usually of part of the lungs, associated with exposure to asbestos.

Term	Definition
MSDs	Musculoskeletal disorders. Injuries to the muscles, joints and tendons of the body, incurred through overuse.
Mycosis	Any disease caused by fungus.
Noise-induced hearing loss	Damage to your ears from equipment at work, eg loud drilling.
PPE	Personal Protective Equipment. It includes safety glasses and hearing protection.
Pneumoconiosis	Lung diseases from inhaling dust over time.
Psittacosis	An infection that birds carry. It can be passed on to humans through contact with faeces, feathers or inhaling cage dust.
RPE	Respiratory protective equipment (a type of **PPE**)
RSI	Repetitive strain injury. Injuries to the muscles, tendons and joints of the body, caused by a movement being repeated.
Silica	Silica occurs as a natural part of many materials used in the construction industry. It may be present in sand, sandstone and granite, clay, shale and slate, chalk, limestone and other rock.
Silicosis	This is a disease caused by breathing in silica dust. The dust causes scarring of the lung tissue, which leads to breathing difficulties.
Tetanus	An acute infectious disease that affects the nervous system, caused by bacteria entering the body through broken skin. All site staff must be vaccinated against tetanus as it can be fatal if a person is unprotected.
Ultraviolet light/radiation	Invisible short-wave length radiation. Sunlight contains UV rays that are responsible for both suntan and – on overexposure – sunburn.

Glossary

Term	Definition
Weil's disease	See *Leptospirosis*.
VWF	Vibration white finger. An ailment caused by the use of hand tools and equipment which transmit vibration to the hand and arm. VWF causes the fingers to become numb and to begin turning white.
WBV	Whole-body vibration. An ailment (usually back pain) caused by machinery vibration passing through the buttocks or the feet.
Work-related back pain	An injury to your back caused by an activity that you do in your job, eg lifting heavy objects.
Work-related stress	Work-related stress is the reaction people have to excessive pressure or other types of demand placed on them at work.
WRULD	Work-related upper limb disorder. Injuries to the muscles, tendons and joints of the body, caused by an activity that you do in your job, eg painting ceilings, which can make your shoulders ache.

This section introduces the benefits of good health practice and why it should always be adopted. It also examines the most common construction processes and identifies the materials that can affect your health.

Why do you need this book?

- ill-health can have a major impact on your life and the lives others

- ill-health problems are often invisible and develop slowly or later in life

- prevention and proper action does make a difference

- the cost of ill-health to you is increasing, in terms of reduced quality of life and lost earnings

- the cost to your employer is increasing, in terms of increased sickness absence and a less healthy workforce

 Health is a state of complete **physical, mental, and social well-being**, not merely the absence of disease or frailty.

- ill-health is a major problem for construction operatives as it can affect their ability to work and have an impact on their whole life

- ill-health continues to kill and maim large numbers of construction operatives

 Ill health should be taken **seriously**. It can be much more serious than an accident as it may be progressive.

- very often there is a delay between exposure to hazardous materials and activities, and the onset of health problems.

1 Introduction to site health

During recent years the construction industry has concentrated on site safety. However, it is important that everyone working in construction also understands the many health hazards that can also be part of work on a construction site but have more insidious and far-reaching effects.

Why bother about site health?

Some interesting facts about occupational health:

- ill-health in UK construction results in an estimated 4 million days off work each year

- each year there are innumerable cases of musculoskeletal disorders at work

- more than 200 incidences of vibration white finger are reported each year

- the incidence rates for hearing loss, spine/back disorders, musculoskeletal disorders, asbestosis, mesothelioma and dermatitis are significantly higher in construction than in any other industry

- it is estimated that asbestos-related diseases kill at least 600 people in construction each year, with this figure expected to continue to rise

- cement is believed to cause up to about half of all cases of occupational dermatitis in the UK.

Take action to avoid being the next statistic

What is occupational health management?

- the management of health at work

 The main aim of occupational health management is to **prevent ill-health** (rather than to cure it).

- employers and managers have a responsibility to manage health risks

 The law says **employers must consider the health risks** that all their operatives are exposed to (including sub-contracted workers) and help reduce the health risks they face at work and that designers **must design out health risks** inherent in the design choices they make.

- good personal health management, working together with your employer, discussing the way you work, developing joint responsibilities and improving co-operation will all result in positive benefits to your health and the health of others.

 Others can advise you or may even be held responsible if they don't act on your behalf but **only you** can minimise the effect on you of poor health management.

1 Introduction to site health

What you can do

- communicate your concerns
- describe the risks as you see them
- make a positive contribution to the debate
- help to redesign the tasks that affect your health and the health of others.

What are the health issues at work?

> To minimise the impact of work on your health and the health of others, you must:
>
> 1 Raise your concerns with your supervisor.
>
> 2 Check with your GP when you suspect you may have symptoms.
>
> 3 Do not wait for increased damage to be certain of your symptoms before going to your GP.

Section 3 explains the specific health hazards faced by construction workers. The principal health issues are split into three separate groups:

1 Biological and chemical.

2 Physical.

3 Psychological.

Biological and chemical health issues

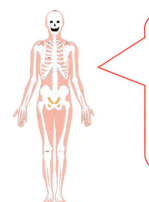

Exposure to many substances used in the construction industry can result in a variety of health problems. These can include dermatitis (skin rashes), breathlessness and burns. There are also long-term effects which can prevent you from continuing work, for example, allergic asthma and skin problems. Some can also contribute to fatal illnesses, including nasal, stomach and lung cancer.

The Control of Substances Hazardous to Health Regulations (COSHH) were devised to ensure that employers consider those hazardous products their employees are using and where necessary implement suitable control measures to eliminate or reduce exposure.

Start by assuming all materials have a hazardous content.

The Chemical (Hazard, Information and Packaging Supply) Regulations (CHIPS) require suppliers to provide safe use labelling and information with and about chemical products.

Be aware that the COSSH issues for initial use/installation/application may be very different from the issues later, when materials are rubbed down, burnt etc.

1 Introduction to site health

Physical health issues

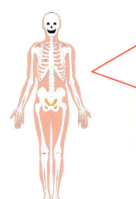

The effects of some physical hazards are often less evident in the short-term and may only become apparent after a number of years have passed. Being aware of the conditions you work in and the result these may have on your health is your first line of defence. **You are your own health and safety officer.**

Work in construction can expose you to most of the currently recognised physical hazards:

- the most common construction health problems such as back pain are triggered by handling, lifting and carrying

- there is also a wide range of skin problems

- noise and vibration are also common health hazards

- heat exposure is not generally a problem in the UK, but work in confined spaces (for example tunnelling work) can be both hot and humid

- radiation is widely used in on-site radiography and in laser equipment. Ultraviolet radiation is not only present naturally, but is produced by electric arc welding

- dust and fumes can be a serious issue in construction and can be hazardous to your lungs, affecting your breathing.

Psychological health issues

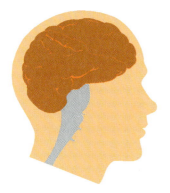

> *Emotional stress is the main problem when discussing psychological ill-health in construction. It can affect us all in one form or another. Awareness raising and positive action can help reduce stress levels in the workplace.*

Stress is the reaction people have to excessive pressure or other types of demand placed on them. Pressure in itself is not necessarily bad and many people thrive on it. It is when pressure is experienced as excessive by an individual that ill-health can result.

By the employer and the employee working together, problems regarding stress can be resolved.

 You are your own health and safety officer. You are responsible for your body, and if you don't look after it will not last you into your old age.

1 Introduction to site health

Materials found on-site and their health implications

If you are in contact with any of the hazardous substances listed below, and you do not take adequate precautions, it is likely that damage to your health will occur.

 You can protect your health by having a pre-planned and disciplined approach.

Table 1.1   *Materials found on site and their health implications*

Materials and substances and where they are used	Health issues arising from using these materials
Asbestos, eg insulation board, ceiling tiles and pipe lagging	Pneumoconiosis, asbestosis, mesothelioma, lung cancer
Carcinogenic materials, eg asbestos, mineral oils, lubricants, and PCBs (found in electrics)	Cancer, mesothelioma
Corrosive materials, eg concrete, brick, acid and wood dust	Cancer of the nasal tract, chemical burns
Skin sensitisers, irritants, eg bitumen, acids, alkalis and cement	Dermatitis
Contaminated land and materials, eg old buildings, redundant gas works, contaminated soils	Anthrax, tetanus, aspergillosis, psittacosis, poisoning, mycosis
Respiratory irritants, eg adhesives, bitumen, solvents	Asthma
Sewage and contaminated water	Leptospirosis (Weil's disease), hepatitis, viral infections

Table 1.1 *Materials found on site and their health implications (continued)*

Silica-based products, eg granite kerbs, masonry, blockwork, fine aggregates	Silicosis
Lead, arsenic, solvents and PCBs found in redundant electrical apparatus	Systemic poisoning
Compressed air in sewers and tunnels	Decompression illness
Direct sunlight, eg when working outside, especially in unshaded areas such as on highways and when doing roofing work	Sunburn, skin cancer
Environments with limited lighting, eg tunnelling	Vision problems
Hot environments, eg when roofing and using hot materials	Heat exhaustion, heat cramps, heat rash, heat stroke
Ionizing radiation, eg welding	Radiation sickness, cancer and eye injuries
Lifting, carrying or moving heavy tools or materials	Work-related back pain and upper limb disorders
Noisy environments	Noise-induced hearing loss
Vibratory tools and equipment	A family of vibration-induced health problems including: – vibration white finger – hand-arm vibration syndrome – whole-body vibration syndrome – noise health issues.

2 Site health – the basics

There are many things that **you can do** to keep yourself in the best health possible. However, it also helps if you know what you can do if you have a health problem at work.

This section gives you some straightforward tips on what you should do **before you start on site**, and what you should do **as soon as you get started on a new site**.

Before you start work on a site, you should:

Register with a local doctor

 It is estimated that nearly 75 per cent of construction workers **do not** have a GP.

The normal practice is for you to be registered with a general practitioner (GP), who is local to where you are living. Although you might move around the country a lot because of your job, it is good practice for you to be registered locally with a doctor. To do this, you should take your medical card to your nearest doctor's surgery and ask if you can register.

 You can approach any practice to ask to be registered there, although doctors are not under any obligation to accept you.

 To find a local GP call 0845 4647 or visit the NHS website: <http://www.nhs.uk>

Don't wait until you are ill to register with a doctor. If you don't have your medical card, speak to the surgery and they will be able to give you a form to fill in.

Alternatively if you need to see a doctor and are not registered with a local GP:

- contact the nearest surgery and ask for treatment
- most town centres have NHS walk-in centres that provide treatment for minor injuries and illnesses and which are open seven days a week. You don't need an appointment and will be seen by an experienced NHS nurse
- you can also consult chemists/pharmacists about medicines and dressings.

For further advice and information about your health, call NHS Direct on 0845 4647 (calls are charged at local rate) 24-hours a day or visit the website: <http://www.nhsdirect.nhs.uk/>.

Some sites will have arrangements with a local GP or will even have occupational health staff on site. Find out from your supervisor at your site induction.

2 Site health – the basics

As soon as **you start work** on a site, you should:

Attend a site induction

Your employer (or, by agreement, the principal contractor on a notifiable project) must provide you with a site induction. It is a legal obligation of employers to provide formal health and safety training.

Site inductions are where you find out about your site and the health and safety issues of which you need to be aware. Part of the induction will probably involve some practical training on the use of tools and equipment. There are also specific restrictions if you are under the age of eighteen.

Your health, your life. Where you have a health issue that may affect you or your work colleagues, make sure your supervisors are aware of your condition.

Remember it is your health and so your duty to ask about training if you are unsure about any of the health issues discussed in this guide.

If you have not been trained you should not be doing the job.

Find out about the health hazards on site

A hazard is something with the potential to cause harm. Harm could include:

- ill-health

- physical injury

- damage to plant, equipment, property or the environment.

 A hazard can include substances or machines, methods of work and other aspects of work organisation.

All health hazards require good management, regardless of whether they result in a minor irritation or major long-term illness.

 Remember – there are **no** "acceptable" health hazards.

With the majority of work-related ill-health, by the time you start to feel ill, the damage has been done. It is therefore important to know the hazards you could face at work, and be prepared to take control and prevent them affecting your health.

 Until recently the effect of construction health issues on long-term health was either not known or considered too far in the future to be of concern. As people now live longer and expect a better quality of life, the effect of work-related ill-health is no longer simply accepted as bad luck.

2 Site health – the basics

Section 3 of this guide tells you about some of the main health hazards you could be exposed to while working on site.

The topics covered include:

3.1 – Hazards that can affect your muscles and joints

3.2 – Hazards that can affect your skin

3.3 – Hazards that can affect your breathing

3.4 – Hazards that can affect your eyes and vision

3.5 – Hazards that can affect your ears and hearing

3.6 – How the working environment can affect your health

3.7 – How stress at work can affect your health

3.8 – How your life outside work can affect your health at work

 The key to better health on site is to work with your employer to:

- remove and reduce the risks as far as possible
- understand about the hazards you face
- control the risk of those hazards affecting you or others.

 Do not put yourself at risk even if others are in danger.

The identification of health hazards and the carrying out of risk assessments and preparing control measures and method statements should be done before you or anyone else starts work on site.

 Hazard is the potential for harm and trisk is a function of **severity of harm** and **likelihood of occurrence**, and the **numbers of people exposed**.

Your employer has a duty to protect your health and safety at work, and you have a duty to protect your own health and safety.

Most of the risks applicable to health that are found on site have been well-defined in terms of health and safety legislation.

Information on risk should be made available to all operatives before they start work.

As well as information on risk:

- site supervisors should provide short training sessions or tool box talks to operatives before starting on a job
- operatives should ask their supervisors about the hazards they face at work.

 Positive action by individual operatives, supervisors and their managers will help improve general awareness of the hazards and risks.

Use personal protective equipment (PPE)

Where health hazards have been identified and cannot be avoided there is a range of suitable protective equipment (PPE) that you may need to wear to protect yourself from ill-health and injury. PPE can also identify you to others due to its high visibility nature, which also improves safety.

2 Site health – the basics

It is law that employers **must** provide **suitable** PPE to their operatives when they are exposed to health and safety risks while at work (unless the risks are already controlled). In the UK, the appropriate PPE that you use must have a **CE mark** on it.

If you are **self-employed** then you also have a duty to **buy** and **use** suitable PPE if there are risks to your health and safety that cannot be controlled through other measures.

The one exception to this is for those who are classified for tax reasons as **self-employed** but otherwise have an **employee-employer relationship**. In these cases it is the duty of the employer to provide PPE.

All workers need to have and use the correct PPE – make sure you are using the correct PPE for the task.

Find out about first aiders/appointed persons on site

Your employer must tell you where you can find first aid equipment and registered first aiders. This may be done as part of your training or through notices posted up in your workplace.

All sites (except perhaps very small ones) will have first aid and qualified first aiders/appointed persons on site. First aiders/appointed persons have undertaken training and obtained a qualification approved by the government.

Make sure you know how to get **first aid** when you start on a new site – ask your supervisor if you are unsure.

If you are a self-employed worker, you must ensure that you have adequate facilities to provide first aid to yourself while at work. You must make an assessment of the hazards and risks in your workplace and put together an appropriate level of first aid provision.

 If you are self employed, you **must** find out if there are any special task-related measures you need to take to protect your health.

Find out about occupational health staff on site

On larger sites there are often a range of health professionals who can give you advice, if:

- you have a question about a work related health issue
- you are concerned that you may have the signs of a work-related health problem.

 Any questions you ask will be **treated as confidential**.

You should find out about the health professionals at your site who are available for you to talk to.

2 Site health – the basics

Depending on the size and nature of your site the following health professionals are able to advise you:

- occupational health doctors and nurses – can advise regarding work-related health problems and give you health checks
- occupational hygienists – can assess the workplace for chemical, biological and physical health hazards
- ergonomists/human factors experts – can provide practical advice regarding the design of work tasks to prevent and control work-related health problems eg suitability of equipment, task design and suitability of PPE
- specialist engineering professionals – can provide advice regarding noise, vibration etc in the working environment.

Report any ill health (three days or more off work) and accidents as soon as they happen, so that a report can be made under RIDDOR

- report all ill-health and accidents to your employer immediately
- report unsafe practices, workplaces and hazards to your supervisor or line manager **straight away**
- be clear about the problem you are reporting.

In the event of an accident or someone feeling ill:

- send for a first aider and/or doctor or ambulance
- send for a supervisor or line manager
- if it is safe to do so, isolate the hazard
- make sure that you and your workmates are safe.

If you have suffered ill-health or an accident at work you need to:

- report accidents to safety reps, write them in the accident book and tell your supervisor

- take note of the advice given by your safety rep about the hazards in your workplace and how you might avoid them

- respond to your safety rep's requests for information, by filling in surveys and attending meetings on both health and safety.

One of the roles of a trade union safety representative or staff safety representative is to check and follow up on management's action over health, safety and environmental issues in the workplace.

Their most important role is to represent workers' views to management.

If you wish to claim compensation against your employer or a third party, contact your local trade union safety representative or staff safety representative. Contact details can be found in Section 4.2.

To claim successfully, any work-related injury you sustain must have occurred because of wrong doing on the part of someone other than yourself.

Even then you do not have an automatic right to damages as compensation for your injury.

2 Site health – the basics

Take action if you have a health problem:

Action	Person to take action
Recognise the health problem	You or your health and safety advisor, line manager, nurse or doctor
Diagnose the problem	Your nurse, doctor
Get treatment	Your nurse, doctor
Discover the cause	You, or your health and safety advisor, hygienist nurse, doctor
Monitor, eliminate and control the cause	Your health and safety advisor, line manager, hygienist, ergonomist
Monitor health	Your nurse, doctor, specialist advisor

Use any health screening that is available to you

Some companies use health screening as a way of identifying potential ill-health problems at an early stage. Make sure you take advantage of any facilities that are available to keep you in good health.

What is health screening?

Screening means carrying out a simple test to check the state of your health. It is done:

- to protect the health of the prospective employee

- to protect the health and safety of others

- to protect the process (or product)

- to enable management to make an informed decision about the suitability of the worker for the task.

 It is important to be screened even if you do not think you have a health problem or that you are at risk of ill-health. Screening will provide a benchmark of your health at a point in time.

 Screening has the potential to save lives and improve quality of life through early diagnosis of serious conditions.

Screening can reduce the risk of developing a condition but it cannot offer a guarantee of protection – that's up to you, working safely with your employer.

What health screening can do for you

"Over a few years, my hearing got gradually worse and worse and in the end I could hardly hear anything at all. I thought I was **deaf**. A couple of years passed when the company I worked for organised these medicals for us. I went along to get my blood pressure checked out and that. I couldn't believe it when the nurse syringed my ears – I could hear properly again after all that time!"

 "I didn't think I had anything wrong with me – I thought I was pretty fit and healthy. I just had the odd ache and pain but put it down to my age. I went for a health screen when I started on this big site and was told that I had **hand-arm vibration syndrome**. I was pretty worried that they'd take me off the job after that. However, the nurse that screened me wrote these recommendations for my supervisor and my work was changed a bit to stop me from working with certain tools. I did notice a big difference and stopped getting achy hands and arms. I've been given new jobs to do as well, which I'm enjoying. Now I **know** I'm healthy…"

 If you are aware, or are made aware by your GP, of a **health problem** that might affect:

- **you, or**
- **your work, or**
- **the colleagues you work with**

then you should see the site nurse or your own GP to make sure you are fit for work. They will advise you what to do next.

It is **your responsibility** to make your immediate supervisors aware of any health problem and then to negotiate your personal needs and the needs of your employer through them.

You will be **better-respected** and less likely to cause trouble for yourself, your workmates or employer if you do this as soon as you are aware of a problem.

 A real life example of health screening

At a large site in 2003, over 3000 people who were to be employed in "safety critical" jobs were screened before they started work. One in three of these workers were referred on for treatment for a medical condition which was identified by the screening. Most of these were eyesight problems or high blood pressure, which were quickly treated allowing the worker to start without further delay. Three of the 3000 workers were found to have a medical condition serious enough to prevent them from starting work on the site and were referred for further treatment.

And finally – remember...

- positive action by you is the key to taking care of yourself

- information on the major risks, together with advice, are covered in Section 3 of this guide

- if you wish to find more detailed information on any of the issues mentioned in this section, the references at the end of this book are a good place to start looking.

3 Health issues at work

Overview

This section describes the hazards and factors that can affect your health at work.

Each health issue is described and details are given about:

- when you are most at risk
- the signs that your health is being affected
- what you can do to prevent health problems occurring.

Contents

INTRODUCTION

Many injuries among construction workers are short-term sprains and strains of the muscles. Construction work can also cause long-term damage to the joints, bones, and nerves. These injuries often occur through constant use, resulting in wear and tear on the body. These injuries are known as **musculoskeletal disorders (MSDs)**.

- MSDs affect the muscles, joints, and tendons of the body
- the most common areas of the body to be affected are the back, knees, neck and shoulders
- you may experience symptoms of pain, aching, and discomfort.

 By the time you start to become aware of the symptoms, your body could be permanently damaged.

 Such injuries may also be called:

- repetitive strain injuries (RSI)
- work related upper limb disorders (WRULDs)
- work related back pain.

MSDs can stop you from being able to work

- you may not be able to manage physically at work which may affect your future employment options.

MSDs can stop you from living a normal life outside work

- you may not be able to continue everyday activities, including simple things such as:
  - ❏ gardening and DIY
  - ❏ playing sport
  - ❏ doing up your shoelaces/buttons.

3.1 Hazards that can affect your muscles and joints

 Remember: MSDs can ruin your life.

MSDs have many different causes. This section highlights some of the key hazards that can cause or worsen MSDs, and when you are at risk. This section covers:

- manual handling
- working with vibration:
    - hand-arm vibration (HAVS)
    - whole-body vibration (WBV).

 Risk from MSDs

Laying kerbs by hand requires workers to carry heavy, awkward loads, and over time this will lead to damage to the back and other parts of the body. Mechanical lifting aids, such as vacuum lifters, are available and allow heavy kerbs and similar products to be lifted into place by machine. Always use mechanical lifting aids whenever possible. Don't ignore discomfort – instead, consult your supervisor.

MANUAL HANDLING

At work there are many times when you may be:

- lifting
- carrying
- moving
- lowering
- putting down
- holding
- pushing
- pulling
- restraining.

...equipment, products or tools.

This is known as **manual handling** and means dealing with materials or equipment using hand or body strength.

The equipment or materials being handled can be of varying size, shape and weight. So they need to be handled in different ways according to the task. Examples of manual handling include:

- shovelling concrete
- pulling a lever
- carrying a ladder
- operating a power tool.

When you are most at risk of injury

You are at risk of injury when your task involves:

- repetitive lifting
- heavy lifting
- bending and twisting of the body
- frequent repetition of an action
- an uncomfortable working position
- exerting too much force
- working too long without breaks
- working in an uncomfortable environment (eg too hot, too cold).

3.1 Hazards that can affect your muscles and joints

 You can be causing gradual damage to your body, even if it doesn't hurt now. It will be painful in the future.

You are more at risk of injury when:

- it's a cold day – your body is more prone to injury when your muscles etc are cold

- you are starting a new job

- you have been away from work for two weeks or more.

You are also at risk when:

- you are working under pressure, eg high job demands, time pressures and lack of control

- you do not report any symptoms immediately.

 Don't forget that manual handling can worsen existing injury. If you have a sports injury, eg from football, then you may have to alter your working practice until you are fully recovered.

 Supervisors/managers:
- have you done your risk assessments?
- have suitable plans been made?
- have the right people been informed?
- has any necessary training been given?
- has any necessary equipment been procured?
- are you now confident people can and will work safely?

Signs that tell you manual handling is affecting your health

- pain, tingling or numbness
- shooting or stabbing pains
- swelling or inflammation
- burning sensation
- stiffness
- aching.

- you may start by having a bit of soreness in the affected areas of the body, especially at the end of a day

- as time goes on, the aching usually worsens, especially if you are still doing the same tasks. The pain usually happens more often too

- this discomfort gradually gets worse until you cannot continue your job.

 Many people will experience aches and pains when they start new work that involves physical effort. However, these aches and pains should not continue – do check them out with your GP or your site occupational health staff if they persist.

What you can do to prevent manual handling affecting your health

DO

- query the need for manual handling with your supervisor
  - ❑ avoid manual handling whenever possible
- use any manual handling equipment that is required for the job eg:
  - ❑ hoists
  - ❑ pallet trucks
  - ❑ mechanical lifting aids.

3.1 Hazards that can affect your muscles and joints

- make sure your body is warm before you lift anything – this reduces your risk of injury

- take a moment to plan for your lift

  - ❑ get help if you think you might need it

  - ❑ clear your path.

 Injuries can last forever and be extremely disabling.

- use good manual handling techniques

  - ❑ when lifting from a low level, bend your knees, not your back

  - ❑ keep your back straight when lifting

  - ❑ get a firm grip on the load

  - ❑ keep the load close to your body

  - ❑ when turning, move your feet, do not twist your body

  - ❑ carry the load for short distances only.

- split the load into smaller/lighter loads wherever possible

- follow advice that is provided to make the job safer

- take regular breaks, or rotate tasks, if the work is repetitive

- report any symptoms immediately

- consult with your supervisor or your health and safety officer on any difficulties or discomfort you may be experiencing.

 Remember that you may be injuring yourself even though the pain has not yet started.

Real life: living with backpain

"When I was about 17, I started getting some back pain after working on a building site. I remember having a heavy week, putting up lots of roof felt, and being up on the roof a lot. We wanted to get finished early so we worked every day straight through, without a break. Time went on and the twinges got gradually worse. I was young, stupid and carried on as normal. It started to feel like I'd got a knife stuck down my back. It was just about bearable in the day time, with mind over matter. At night though, I couldn't sleep. I eventually went to the doctor but by then it was too late really. They couldn't do much except put me on painkillers and tell me to rest. I did have some physiotherapy but my back was in such a mess it didn't really make any difference. I couldn't do any sport anymore, and the pain was getting worse and worse. I'm now on so many painkillers that I've ended up on more tablets for other things such as constipation. I've had to stop work and spend my time lying in agony. I feel useless. No-one can tell me how to improve things, only that it is more likely that I will get worse than better. I am only 28 years old. I wish I'd been sensible and gone to the doctor as soon as I started getting a pain. Then something could have been done about it."

This ex-site operative was involved in awkward, repetitive work, and didn't take regular breaks from the task. He ignored his backache until it became too painful to cope with. By then his back was severely damaged.

3.1 Hazards that can affect your muscles and joints

DO NOT:

- twist or stoop the body when you lift

- hold the load away from the body

- carry loads over a long distance

- move up and down different levels carrying loads

- carry loads that make you off balance

- move a heavy or awkward load

- carry loads when you can't see where you're going

- continue to work in discomfort – develop improvements with your supervisor

- wait for increased damage to be certain of your symptoms before going to your GP.

For information on the limits of how much weight you are allowed to carry, have a look at the HSE's *Manual handling assessment charts* (see Section 4.2 for further details). They may also be downloaded from <www.hse.gov.uk/msd>.

WORKING WITH VIBRATION

Vibration can affect your body while you are at work. This is especially likely when you are using high-vibration tools for long periods. Such tasks can result in **hand-arm vibration syndrome** (HAVS).

Vibration can also affect your health if it is present in the environment in which you are working, eg when driving a vehicle. Working in vibrating environments can result in **whole-body vibration** (WBV).

Vibration is a problem because it can permanently alter the flow of blood around the body. If areas of your body are not receiving a good blood supply, they can be damaged. **Any part of your body can be affected** by vibration but some of the most common parts to be affected include the **arms, hands, and back**.

 By the time you start to become aware of the symptoms, your body could be permanently damaged.

When you are most at risk of injury

Hand arm vibration syndrome (HAVS):

- if you regularly use high-vibration hand-held tools then you are at risk, for example:
    - ❏ manual grinding tools
    - ❏ riveting tools
    - ❏ chainsaws and similar woodworking machines
    - ❏ power hammer such as caulking and chipping hammers or concrete breakers
    - ❏ percussive drills.

 You are at high risk of HAVS if you use high-vibration tools.

- the risk increases when:
    - ❏ you use the equipment for a long duration
    - ❏ the vibration levels from the equipment are high
    - ❏ you get cold and wet when using the equipment
    - ❏ you have to grip the equipment tightly
    - ❏ it is awkward for you to use the equipment
    - ❏ equipment is poorly maintained.

3.1 Hazards that can affect your muscles and joints

 If you are using more than one high-vibration hand-held tool, remember to add up the time you spend using all of them, eg riveter, chainsaw, percussive drill etc.

 Information on the amount of vibration that a tool emits will be available. Ask your supervisor or health and safety officer for the safe use (trigger) time for a tool before you use it. Suppliers and manufacturers will also have this information.

Whole-body vibration (WBV):

You are at risk of WBV if your job involves:

- driving construction vehicles, eg mobile machines, fork lift trucks and quarrying or earth-moving machinery.

Regular long-term exposure to WBV is associated with back pain and other factors such as poor posture and heavy lifting.

 Pay attention to your seating when you are driving. A good seating position and feeling comfortable are important for reducing WBV.

Signs that tell you HAVS is affecting your health

- if you start having pins and needles and/or numbness in your fingers, especially at the end of a day

- if you have problems with your fingers during cold weather. You might find that the tips of your fingers go white, the fingers become pale and you lose the feeling in your fingers. When returning to a warm environment after being out in the cold, your hands may flush red and throb painfully.

Vibration can affect your nerves, muscles, bones and joints as well as your blood circulation.

As time goes on, the condition usually worsens, especially if you are still doing the same high vibration tasks. This usually means that the pins and needles/numbness happen more often and more severely, and not just after a day's work or when working in the cold. Attacks may start to occur when working in warm environments.

- because you cannot feel your fingers very well, you lose your nimble fingers and may not be able to do many tasks, putting your job at risk

- in worst case scenarios your hands may become like sausages, poorly co-ordinated, clumsy, with regular frequent attacks of white finger, causing great discomfort.

 Vibration white finger (VWF)

VWF is a typical example of HAVS. VWF is provoked by the hands being cold. Once re-warmed some discomfort may occur as the blood circulation returns.

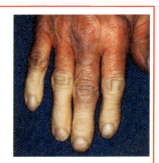

3.1 Hazards that can affect your muscles and joints

What you can do to prevent vibration affecting your health

DO:

- review the need for vibration exposure with your supervisor

- tell supervisors/health and safety officers if equipment is faulty – faults with equipment often result in greater levels of vibration, so get them fixed

- take **regular breaks**, or **rotate tasks**, if the work is repetitive

- use **low vibration tools** or tools with low vibration handles wherever possible.

Managers:

- have you tried to eliminate or mitigate the risk?

- have you got a policy for tool selection?

- have you got low vibration tools on site?

- if so, have you informed the workforce?

- where there is vibration, there is often noise – have you got noise protection in place?

- try to keep the hands and body warm. This helps to keep a good flow of blood moving around the body, which cuts down the chances of injury. You can help to keep yourself warm by:

  - wearing gloves (for warmth and protection)

  - using heating pads

  - using tools with heated handles

  - wearing warm weatherproof clothing

  - doing warm up exercises before starting the job

  - using a warm, sheltered area when you take breaks

  - exercising and rubbing (massaging) the hands and fingers during these breaks.

- avoid smoking – smoking has an effect on your blood flow so try to give up smoking as far as possible as this effect combined with doses of vibration is not healthy

- consult with your supervisor or health and safety officer on any difficulties or discomfort you may be experiencing

- report any ill-health to your employer – this is your own responsibility.

 It's your health and quality of life that suffers. Once you've got HAVS, you're stuck with it.

DO NOT:

- use vibrating tools and equipment for long periods – take regular breaks

- use vibrating tools for full shifts – your supervisor must inform you of the safe trigger time for each tool

- continue to work in discomfort – develop improvements with your supervisor

- wait for increased damage to be certain of your symptoms before going to your GP – talk to your supervisor, visit your doctor and explain about the work you do.

 Working in cold conditions can make you more likely to be affected by vibration. See Section 3.6 for further information on working at environmental extremes.

3.1 Hazards that can affect your muscles and joints

Real life: living with HAVS

"I never thought that I had got a problem that was caused by the tools at work. I had been having cold fingers in the winter for years. I thought it was just me, or something to do with my age.

Two of the fingers on my right hand started going numb sometimes when I had used the angle cutter. It got gradually worse and now my hands are sore and they ache all over. Thinking back, I didn't really pay attention to any of that health and safety stuff about how you shouldn't use those sorts of tools for long periods. As if that wasn't enough, I can't use my hands properly anymore as I've lost all my feeling – it's difficult to do even simple things like buttoning up my shirt and tying my shoelaces. It makes life awkward and I have to get people to help me which is embarrassing."

This site operative is still at work although is now unable to do his original job. He is now unable to do many normal activities that people take for granted. He also lives with constant pain in his hands and arms.

INTRODUCTION

When working in construction, there are many times when your skin may be exposed to harmful substances, activities or environments.

The biggest problems are:

- work related skin complaints
  - ❑ eg dermatitis, burns from working with cement
- diseases caused by exposure to ultraviolet light
  - ❑ eg skin cancer, sunburn
- diseases caused by exposure to radiation
  - ❑ eg burns, cancer
- infections spread by contact with the skin
  - ❑ eg tetanus, Weil's disease.

3.2 Hazards that can affect your skin

WORK RELATED SKIN COMPLAINTS

These are caused by exposure to various substances, activities or environments that contain irritants. The skin on the **hands** is the most commonly affected area but skin on the **face, neck, chest, arms**, and **legs** can also be affected.

There are two types of dermatitis:

1   **Irritant contact dermatitis:** caused by contact with substances which cause irritation at the site of exposure. If exposure is prevented, then the symptoms will clear up.

2   **Allergic contact dermatitis:** a more serious complaint where the body becomes sensitised to the substance which triggers the dermatitis, and the disease will be triggered by very small amounts of the substance. Sufferers from allergic dermatitis caused by cement will often have to leave the industry.

You are at risk of work related skin complaints when

- working with concrete, oils or solvents, plaster, epoxy resins, insulation, paint, machinery – any job where you are exposed to dust, chemicals, or contaminants

- working in trenches – you may be exposed to infected water and contaminants resulting in disease or skin infections

- cleaning or collecting waste – you may be exposed to contaminants.

You are always more at risk if your skin is exposed to weather elements (eg sun, wind, rain). This is because these conditions damage the thin top layer of skin, leaving the underlying skin more prone to access by irritants such as cement, mortar, oils, detergents etc.

These all depend on the strength of the contaminant, the length of time the skin is in contact and the sensitivity of your skin.

Concrete

Concrete dermatitis is a common and serious skin condition. It is due to contact with concrete, often

resulting when PPE has not been used correctly. It can be extremely painful, resulting in burning, cracked, raw skin.

Signs that tell you there is a problem

- if your skin appears dry, red, sore and sometimes itchy. This reduces the skin's ability to be able to cope with the effects of chemicals, dust etc.

- if, even if only one area of the body was originally affected, you find that the dermatitis spreads to other areas of the body

- if the outer layer of skin shrinks and becomes brittle and cracks

- if the cracks get deeper and start to bleed, it means that dust, chemicals and bacteria can get into the sensitive, underlying tissue, and your skin will get very sore and inflamed

- if the cracks become deeper this can be a way of substances reaching the internal organs, which can cause serious damage to your health.

3.2 Hazards that can affect your skin

 Contact

Contact dermatitis is caused by contact with irritants such as acids, solvents, detergents and even sugar, flour 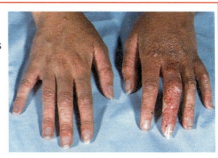 and soil. It is usually contracted by not using PPE correctly and it results in painful, blistered skin.

What you can do to prevent work related skin complaints

DO:

- consider specifying alternative materials and methods to reduce the risk of skin damage

- use protection

  - ❑ protect the skin by avoiding contact with the irritant by wearing gloves and other protective clothing, eg suitable overalls

  - ❑ replace PPE that is worn, torn or damaged and keep PPE clean

  - ❑ make sure that the gloves you wear are the right length, size and material, eg waterproof – if you have any queries, check with your health and safety officer

  - ❑ if you already have any cuts or abrasions, make sure they are covered with waterproof dressings before you start working

- avoid touching irritants

❑ wherever possible, use tools or equipment to handle the substance.

 Liquid, high alkaline cements and grouts can result in painful problems.

● keep clean – try to keep the workplace and equipment, as well as your PPE and clothing, as clean as possible from contamination. Irritants and bacteria are easily spread by touch.

 Wash your hands after handling cement, chemicals, solvents and oils. Unwashed hands spread infection. See Section 3.6 about the other important reasons for washing your hands.

● wash any contaminated skin immediately. Use specialist skin cleaners to remove oil and grease. Do not just use "any old thing" that's close to hand – using cleaners designed for tools, eg white spirit and turpentine, can dry out the skin and cause further problems

❑ in winter months, when your skin is exposed to windy, wet and cold weather, use a barrier cream. This protects the top layer of your skin when it is at risk of being removed by the harsh weather

❑ when you finish work, after you have thoroughly washed your skin, use emollient and moisturising skin creams. This is to re-moisturise your skin and keep it supple, which prevents it from getting damaged.

3.2 Hazards that can affect your skin

Look after your hands

Emollient cream that you can put on your hands after you have washed and cleaned them can make a big difference. Doing this prevents the skin from becoming sore  and cracked, averting dermatitis, as well as infection through the skin. These creams are readily available from any chemist/pharmacist. Ask your health and safety officer to supply some on site.

- protect your health by improving working methods with your supervisor

- tell your supervisors, or the site health and safety officer, **immediately** if you have any of the signs and symptoms mentioned above. **This is your responsibility.**

DO NOT:

- wait for damage to be certain of your symptoms before going to your GP.

ULTRAVIOLET RADIATION

Ultraviolet radiation (UV) is given off by the sun. You are exposed to ultraviolet radiation when working in direct or reflected sunlight (usually when working outside).

You are at risk of diseases caused by UV light when

- you are working outside in the sun you and exposed to ultra violet radiation, which can cause skin problems. Reflected sunlight may also affect your vision, due to glare. You may feel unwell, eg sunstroke – sickness and headaches

    ❑ it can result in skin cancer after significant exposure

- you are using particular products – some chemicals can increase your skin's sensitivity to the sun, eg wood preservatives. This is because sunburned skin is already damaged and chemicals can easily get through the top layer to the more sensitive underlying tissue

- you are doing any electric arc welding – you are also exposing yourself to other forms of radiation at the same time.

All of these factors depend on the strength of the contaminant, the length of time the skin is exposed and the sensitivity of your skin.

Signs that tell you there is a problem

- if the skin on your body that has been exposed to the sun becomes red and feels tender, and if you develop blisters

- if you experience the feelings associated with heat stress (see Section 3.6), eg nausea, dizziness, headache etc

- if you develop irregularities with moles or spots on your skin. These changes may indicate serious damage to your skin cells from direct exposure to sunlight. This may be a sign that you are developing skin cancer, which can result in death if not diagnosed.

If you do notice changes to moles or spots on your skin you should get them checked out with a doctor.

3.2 Hazards that can affect your skin

What you can do to prevent UV damage

DO:

- protect your skin from the sun – even if it's a cloudy day, you are still at risk. You can protect your skin by:
  - ❑ covering it up, eg wearing a hat, long sleeved top, and trousers
  - ❑ using sun cream with a high sun protection factor (SPF)
  - ❑ wearing sunglasses to protect your eyes
- avoid working in direct sunlight for long periods, especially when the sun is at its strongest, typically between the hours of 11.00 am and 2.00 pm
- protect your health by improving working methods with your supervisor
- minimise damage to your health by seeking help for problems as soon as you are aware of any
- tell your supervisors, or the site health and safety officer, if you have any concerns about your health. This is **your responsibility.**

 Sunburn can be very damaging to your health

Skin that is unprotected by clothing or sun cream from the sun's ultraviolet light can blister and be very painful.

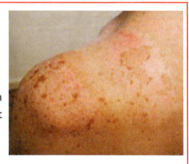

DO NOT:

- wait for increased damage to be certain of your symptoms before going to your GP.

RADIATION

Radiation is energy that is transmitted, emitted or absorbed in the form of particles or waves. The larger the amount of radiation and the greater the length of time you are exposed, the more chance you have of developing radiation disease. Working in areas where you are exposed to high amounts of radiation can eventually kill you.

You are at risk of diseases caused by radiation when

- working with external radiation sources, eg radiography

- working near external radiation sources, eg if you go into unauthorised high risk areas.

Signs that tell you there is a problem

There are few early warning signs of illness caused by radiation

- as the illness gets worse you may suffer from:
  - ❑ feelings of nausea
  - ❑ vomiting and diarrhoea
  - ❑ exhaustion and fainting
  - ❑ redness, tenderness and swelling of areas of your body
  - ❑ hair loss
  - ❑ ulcers in your mouth, throat and intestines
  - ❑ bleeding from your nose, mouth and rectum
  - ❑ sore skin, including open sores on the skin surface.

3.2 Hazards that can affect your skin

What you can do to prevent radiation diseases

DO:

- protect your health by improving working methods with your supervisor

- always check with your radiation protection supervisor (RPS) before working near any radioactive sources

- minimise damage to your health by seeking help for problems as soon as you are aware of any

- tell your supervisors (or the site health and safety officer) if you have any concerns about radiation affecting your health. This is **your responsibility.**

DO NOT:

- put yourself in a position where you might be exposed to radiation.

 Never cross a barrier displaying a trefoil. This warning sign is telling you that beyond it is an area of radiation and it is not safe to go past it.

 If you are working in a radiation contaminated area it is law that your employer gives you regular health checks and education.

INFECTIONS CAUGHT THROUGH SKIN CONTACT

These are infections caught through the skin, eg tetanus, Weil's disease. These are often caused by exposure to dirty, infected water or tools.

You are at risk of skin borne infections when

- working anywhere on site, particularly in contaminated land or water, including if you are in contact with sharps and needles

- excavating near ditches or ponds

- working in sewers

- working in waste management

- after heavy rain, eg a thunderstorm, which has created movement of ground water.

Weil's disease
(leptospirosis)

Weil's disease is caused by water that has been infected by bacteria in rats urine. For example, water that has collected in the

bottom of trenches may have become stagnant and infected. To avoid problems, you should wear suitable PPE and wash any skin that has been in contact with the polluted environment. Thousands of people contract the infection every year, and most recover completely with treatment. If untreated you may not survive the organ damage caused by the bacteria. The onset of symptoms is rapid, and in severe cases decline is also very quick.

3.2 Hazards that can affect your skin

Signs that tell you there is a problem

- symptoms may include:
  - ❑ a sudden high temperature
  - ❑ a flu like illness
  - ❑ other symptoms that appear like:
    - joint and muscle pain
    - conjunctivitis
    - eye infection
    - jaundice (yellow skin) – this is a liver problem.

 People with poor kidney function or existing skin conditions such as eczema may be at a higher risk of getting such an infection.

 It is very important to get early medical attention if you have any of these symptoms. This is the only way to prevent serious illness.

What you can do to prevent skin-borne infections

DO:

- keep clean – wash or shower thoroughly after working in stagnant or slow moving water and soil
  - ❑ wash your hands regularly and always wash your hands before eating or drinking
  - ❑ avoid rubbing your face during work as bacteria can spread via your eyes, nose and mouth
  - ❑ wash and clean your clothes and PPE regularly.

- cover up

  - ❑ cover cuts and grazes with waterproof plaster

  - ❑ wear personal protective clothing, eg gloves, waders. This protects your skin from being grazed and bacteria getting into your body

  - ❑ make sure that the gloves you wear are the right length and size – if you have any queries, check with your health and safety officer

  - ❑ keep your PPE clean.

 Dirty gloves can often be worse than wearing no gloves at all.

- protect your health by improving working methods with your supervisor

- go to your doctor or local hospital as soon as you sense a flu-like illness that persists, and you are regularly working in stagnant water, soil, or sewers. Make sure you tell the nurse/doctor about your job

- minimise damage to your health by seeking help for problems as soon as you are aware of any – do not wait for increased damage to be certain of your symptoms before going to your GP

- tell your supervisors (or the site health and safety officer) if you have any concerns about your health. This is **your responsibility.**

 If you get flu-like symptoms that persist and you have been working on a construction site, let you GP know.

3.3 Hazards that can affect your breathing

INTRODUCTION

When working in construction, there are many occasions when you may be at risk of breathing in something that could harm you. There will be similar risks when carrying out intrusive investigations (eg removing ceiling tiles) or testing (eg drilling for samples).

The hazards may include dust or small particles of:

- everyday hazards
    - ❑ wood dust
    - ❑ cement dust
    - ❑ solder flux
    - ❑ glues and resins
    - ❑ mineral fibre dust
    - ❑ lead or isocyanates (found in paints)
    - ❑ bird droppings
- asbestos
- silica.

 The list above shows that common, everyday products can cause problems to your health when breathed in.

When you are at risk of respiratory hazards

- woodwork
- insulating
- demolishing
- cutting MDF
- painting and removal of paint/finisher
- cutting concrete
- mixing concrete
- welding
- maintenance
- grit blasting.

Common hazards to your lungs and breathing

Cutting concrete kerbs and similar products creates large amounts of dust containing silica which can affect your breathing or cause silicosis (see section on silica). Water suppression should be fitted to the cutter to reduce the amount of dust in the air during the job, and a dust mask should be worn along with suitable gloves.

EVERYDAY HAZARDS

Signs that tell you everyday hazards are affecting your health

- you may have bouts of coughing or wheezing and you may find yourself short of breath

- your chest may feel tight and you may find it difficult to breathe

- you may find that your nose is runny or stuffy, and you are sneezing frequently

- you may have watery or itchy eyes and a tickly throat

- you may find that these symptoms happen more and more regularly, even when you are not working with these hazards

- you may find that you become far more sensitive to other factors, which have rarely caused you a problem before. These may include animal fur, smoking, and pollution

- these initial signs may develop into full blown asthma attacks, where breathing is very difficult, the chest feels very tight and you wheeze very badly.

3.3 Hazards that can affect your breathing

Some people may have been working with hazardous products for years and suddenly notice a problem. For other people, the build up of symptoms is more gradual.

Real life: living with occupational asthma

"I left school and became a painter and decorator. In the early 1990s I was an exterior painter and after a while was transferred to a paint spray shop. After about six months I started to get a bit tight in my chest and a bit **wheezy**. I thought it was just because I was a **heavy smoker**. Not long after I started getting wheezy, I was moved to working outside again. That didn't help – it actually made my chest worse.

After another year or so my chest got so bad I **couldn't work** with the paints anymore. It was giving me problems all the time, difficulty breathing, and even talking sometimes. I had to leave all that behind at the age of 33 and I'm trying to **retrain** at the moment. The asthma's got a bit better now, but I can't do any DIY or hands on work anymore. I've become really sensitive to dust of any kind – anything sets me off wheezing. I sound like an old man.

As for a cigarette – that would **kill** me straight away.

If I'd been using safer materials, and been given the right PPE to wear, it would never have become such a problem. It was never really picked up on either, because we never had any health checks. I never thought about wearing a mask either and it could have made all the difference. It was such a gradual thing.

If only I'd **known** what a bit of paint spray could do to me back then."

What you can do to prevent everyday hazards affecting your breathing

DOs:

- know what you're working with

  - ❑ do read the health and safety data sheet

  - ❑ do spot the hazards – speak to your supervisors about anything you are unsure of

- do use protection

  - ❑ use personal respiratory protection to prevent yourself breathing in dust etc, eg face masks (which should be provided by your employer).

 Use appropriate PPE

Use the suitable PPE for the task – different masks serve different purposes. Make sure you keep PPE clean, and follow the instructions for its use. Above all, make sure it fits snugly.

- do replace the filter on face masks at regular intervals as stated in the instructions and check before use

- if the filter appears discoloured, eg brown or red, replace it with a new one

- if it's a disposable face mask, use a new face mask on a daily basis

- do check you've got the right PPE and that it's working properly

3.3 Hazards that can affect your breathing

- do keep clean

    - ❑ try to keep the workplace and equipment, as well as your mask and clothing, as clean as possible from dust and other contaminants

- do give up smoking if possible, as smoking can make symptoms even worse

- do protect your health by improving working methods with your supervisor

- do minimise damage to your health by seeking help for problems as soon as you have any

- do tell your supervisors or the site health and safety officer **immediately** if you have any of the signs and symptoms mentioned above. This is **your responsibility**

- do not wait for increased damage to be certain of your symptoms before going to your GP – remember there is often no cure for sensitivity caused by exposure to everyday hazards.

ASBESTOS

What is asbestos?

Asbestos is the name given to a group of naturally occurring minerals used in certain products including building materials.

 You should be working with asbestos only if you have been **formally trained** to undertake such work.

When you are at risk of developing asbestos-related diseases

- you are at risk when you are exposed to breathing asbestos **fibres**

- the presence of **intact** asbestos materials on site does not mean that there is always a risk to you

- any work that disturbs the asbestos (including **asbestos cement**) may result in asbestos fibres being released into the air. Work may include:

  - ❑ intrusive investigations (such as moving ceiling tiles)

  - ❑ breaking, drilling or cutting

  - ❑ demolition

  - ❑ working with old boilers.

 Occasionally, asbestos might be present without you knowing about it. If you think asbestos might be present in your workplace, do not disturb the suspect material and speak to your supervisor and your health and safety officer **immediately**.

3.3 Hazards that can affect your breathing

You are at risk if you disturb any **asbestos containing materials (ACMs)**

 If you don't know if you're qualified to work with asbestos **then don't**.

The places in buildings where asbestos is likely to be found are:

- ceiling voids and ceiling tiles
- lagging for thermal insulation of pipes and boilers
- fire protection in ducts, firebreaks, panels, partitions, soffit boards, ceiling panels, and around structural work
- insulating boards for fire protection, thermal insulation, partitioning and ducts
- asbestos paper used to back vinyl flooring, or to insulate electrical equipment
- asbestos cement products – used in roofing, wall cladding, guttering, rainwater pipes, and water tanks
- certain textures coatings, ie artex
- bituminised products including flashing and floor tiles.

 If your work involves cutting, drilling, sawing or breaking up materials, your employer will need to be sure that there are no asbestos containing materials present. If there is doubt, your employer will need to arrange for a survey to be carried out.

Signs and symptoms of asbestos-related diseases

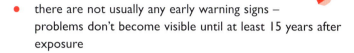

- there are not usually any early warning signs – problems don't become visible until at least 15 years after exposure

- after exposure to asbestos fibres, you may develop difficulties with breathing as well as:

  - ❑ lung cancer

  - ❑ mesothelioma

  - ❑ chronic lung disease (eg asbestosis).

These conditions can kill you

What you can do to prevent asbestos-related diseases

DOs:

- do remember that there is no cure for asbestos related disease

  - ❑ do stop work immediately if you think that there is any asbestos dust or fibre where you are working

- do tell your supervisors (or the site health and safety officer) **immediately** if you think asbestos is present. This is **your responsibility.**

Managers/supervisors:

Ensure that you are aware of the presence of ACMs from the asbestos surveys/register before starting any work on existing structures. If ACMs are present you must:

- redesign the tasks to avoid disturbance of the ACM, or
- carry out the task with a licensed contractor.

3.3 Hazards that can affect your breathing

- do only work with asbestos (and ACMs) if you are qualified to do so

- do make sure that a method statement/safe system of work has been prepared and you are familiar with it

- do remember that there is no cure for most asbestos-related illness, so do stop work and report to your supervisor immediately if there is any asbestos or ACM present.

 Asbestos kills – there is **no cure** for asbestos-related disease.

 Remember – any exposure to asbestos can result in the development of asbestosis for which, as yet, there is **NO CURE.**

DO NOT:

- where asbestos has been removed, do not cross barriers or enter restricted areas

- do not disturb or work with asbestos (and ACMs) unless you are qualified to do so

- do not remove tools or PPE from a contaminated area. Wait until they have been decontaminated

- do not remove PPE while you are still in a contaminated area – wait until you are in another area on site (normally it will be removed on the dirty side of the hygiene facility)

- do not take PPE home to wash. Contaminated masks and filters should be disposed of as asbestos waste

- do not wait for increased damage to be certain of your symptoms before going to your GP.

Real life: living and dying with asbestosis

"It starts off as nothing much really – colds go to your chest more often and you find that even the slightest cold gives you a cough and makes you breathless. As time goes by and the disease progresses you start to get **breathless** when doing any strenuous work, then you start getting breathless just doing ordinary things like going upstairs. Soon **working becomes impossible** and you have to give up. You can't do anything strenuous, no dancing the night away, no heavy gardening or DIY, and no sex.

As it gets worse you are breathless just getting up from a chair and soon **oxygen is needed** – you're tethered like a balloon you can only go where your tube will let you. By now you have resigned yourself to living downstairs, upstairs is a no-go area. You might be getting some industrial injuries benefit but the system is slow and payouts are kept under control so you won't be getting a lot… probably not enough to pay for all your care needs. You can say **goodbye to any money** you might have put by for a rainy day. You'll need help doing the ordinary things in life – cooking, cleaning, even taking a bath or going to the loo are things you will never do on your own again. It all needs to be paid for too. If you haven't moved house by now you probably soon will so that you have some money to fund **your ever increasing need for care** and attention. As time goes on you'll need the oxygen more and more, first 10 hours per day then 12, 15 and eventually 24 hours a day with tubes up your nose and fastened to gas bottles or an **oxygen machine**. As things get worse, other problems start: the **heart failure** that has been building up due to the resistance in your lungs gets worse, you develop swelling in your feet and hands and to add to the problem in your lungs fluid starts to leak from your **weakened blood vessels** into the lung tissue making it harder still for you to breath.

You're on around 30 or so tablets per day with some being taken to counteract the **toxic effects** of others, you have your oxygen and a nebuliser which you use four or five times per day to try to keep the airways open, what you have left of them.

Every so often you are wracked with **uncontrollable fits of coughing and choking** as your lungs spasm in a desperate attempt to get some air. These leave you exhausted, weak and blue around the lips, it might take you **days to recover** from one and you fear that each one will be your last. To make matters worse the coughing also results in incontinence so you need help to get you out of your soiled clothing. **Talking to anyone becomes impossible**, even if you have the breath to speak you are aware that you smell so you stay in your room, besides you can only go where your oxygen goes. You may even need a catheter to control the incontinence resulting in more **embarrassment and pain**."

This ex-operative was exposed to asbestos before the law was improved. He used to paint asbestos insulation which regularly had to be brushed down to a solid surface so that it would take the paint and the risks weren't widely known.

 Work to the rules – they are there to protect **you**.

SILICA

What is silica?

Silica occurs as a natural part of many materials used in the construction industry. It may be present in:

- sand, sandstone and granite
- clay, shale, and slate
- chalk, limestone and other rock.

When you are at risk of developing silica-related diseases

Silica is present in many materials used in the construction industry. However, it is breathing in the silica dust that can affect your health.

These activities may mean you are exposed to high levels of silica dust:

- quarrying
- tunnelling
- grit blasting
- grit or sand blasting
- demolishing
- scabbling, cutting, or drilling any concrete products.

You are also more at risk of silica dust affecting you if you are a smoker.

Think:

- are you being exposed to silica?
- has a risk assessment been carried out?
- has dust monitoring been carried out?

It's **your health** – find out from your supervisor or health and safety officer.

3.3 Hazards that can affect your breathing

Signs and symptoms of silica-related diseases

- there are not usually any early warning signs that you have been exposing yourself to silica – except an **early death**

- problems tend to come on gradually. This is because there is a gradual loss of lung tissue due to damage from the silica. This affects your ability to breathe

- you start by finding yourself **short of breath.** For example, you find it difficult to walk up and down stairs

- your chest may feel tight and you may find it **difficult to breathe**

- the situation gets gradually worse and there's nothing that can be done to help you

- you end up housebound and then confined to bed as you cannot manage to move about

- you end up **dying at an earlier age** due to heart failure.

 The effects of silica are permanent and **cannot be treated**.

 Your doctor may use the following terms when discussing silica-related diseases: silicosis, tuberculosis, fibrosis, cyanosis.

What you can do to prevent silica-related diseases

DOs:

- do make sure you have been told about all the precautions that need to be taken if you are working with materials that contain silica. If you are at all uncertain, stop work and ask your supervisor

- do make sure that all control measures (eg ventilation, extraction) are working properly. Take any problems with these arrangements to your supervisor

- do use personal respiratory protection to prevent yourself breathing in dust etc, eg face masks (which must be provided by your employer)

- do use cutting tools fitted with water suppression to prevent the dust being created, and make sure there is water in the reservoir.

 Make sure your PPE fits properly – have you had specific training on how to use your PPE? If not, ask your supervisor or your health and safety officer.

- do know what you are working with
  - ❑ use a safer product, eg water based paints
  - ❑ read the health and safety data sheet
  - ❑ spot the hazards – speak to your supervisors about anything of which you are unsure

- do replace the filter on face masks at regular intervals as stated in the instructions and check before use
  - ❑ if the filter appears discoloured, eg brown or red, replace it with a new one
  - ❑ if it's a disposable face mask, use a new face mask on a daily basis

- do make sure the PPE is worn properly and that it fits properly – if it doesn't it is not working correctly and you could be putting yourself at risk

- do protect the skin by avoiding contact with silica by wearing gloves and other protective clothing, eg suitable overalls

- do make sure that the PPE you wear is free from contamination, clean and in a good condition

- do avoid creating dust.

3.3 Hazards that can affect your breathing

Managers/supervisors:

- make sure that one piece of PPE is compatible with another
- make sure that PPE fits all operatives correctly.

- do try to keep the workplace and equipment, as well as your mask and clothing, as clean as possible from dust

- do keep dust to lower levels by:
  - ❑ keeping materials dampened down
  - ❑ using hand tools rather than power tools
  - ❑ get rid of dust as you work, and while it's still damp

- do wash your hands and face before eating, drinking or smoking

- wash your hands and face before leaving the site

- do protect your health by improving working methods with your supervisor

Only on the most severe occasions will signs develop straight away. Usually the symptoms develop 10–20 years later. By then **it's too late.**

- do minimise damage to your health by seeking help for problems as soon as you have any

- do tell your supervisors (or the site health and safety officer) **immediately** if you think silica is present and adequate precautions are not being taken. This is **your responsibility.**

DO NOT:

- do not remove PPE while you are still in a contaminated area – wait until you are in another area on site

- do not allow the water reservoir to run dry on water suppressed tools

- do not take PPE home to wash

- do not continue to work if the protective equipment arrangements become faulty, but check with your supervisor

- do not wait for increased damage to be certain of your symptoms before going to your GP.

 Prevention is key – it's too late when you have the signs of illness.

3.4 Hazards that can affect your eyes and vision

INTRODUCTION

There are many hazards in the construction industry that can affect your eyes and your vision:

- nails, tiny pieces of metal, splinters, and cut wire ends can fly in the air

- the mixing of cement, sawing, grinding, and chipping produce dusts and grit. So does heavy machinery moving across a site

- chemicals and welding arc can burn your eyes

- you are also at risk of damaging your eyes when working in bright sunlight – see Section 3.2 (UV radiation)

- your eyes can be strained when working with computer screens (VDUs) for long periods.

 The construction industry has one of the highest rates of industrial eye injuries.

Eye injuries vary widely among the trades and sectors of the construction industry. Common injuries include:

- scratches from a foreign body

- foreign bodies embedded

- chemical splashes

- radiation burns from welding

- burns from looking into fibre optic cables.

When you are at risk of damaging your eyes

You are at risk when you are exposed to possible eye or face injury from:

- flying objects, eg when using a nail gun
- liquid chemicals
- acids and caustic liquids
- chemical gases or vapour
- light radiation, eg from welding, fibre optic cables
- sparks, eg when disc cutting.

Protection against infection

Infection can develop in the eye from irritation, such as getting a small amount of a chemical in the eye. If untreated, some types of eye infections can damage the eye very quickly.

Signs that tell you your eyes have been damaged

- you may have sore, itchy, weeping, or red eyes
- you may have flashes of light in your eyes
- pain in the eye
- it may feel as though something is in the eye (foreign body sensation)
- you may be ultra-sensitive to light
- a grey or white sore may develop on the coloured part (iris) of your eye
- you may have blurred or decreased vision.

3.4 Hazards that can affect your eyes and vision

What you can do to prevent eye problems

DO:

- do make sure you have been told about all the precautions that need to be taken when you are working. If you are at all uncertain, stop work and ask your supervisor

- do make sure that eye protection is available and working properly. Take any problems with these arrangements to your supervisor

- do use the eye protection equipment that is provided in your workplace, eg safety spectacles, goggles, face shields, hoods and helmets

  - ❑ all eye protection should be:

    - − worn correctly according to the instructions (eg over the top of any prescription glasses)

    - − the correct size for you

    - − kept clean

    - − inspected regularly

    - − replace if damaged

    - − worn at ALL times when you are at risk.

 The bother of wearing eye protection is nothing compared to losing your sight.

- do wear eye protection at all times when you are at risk of injury, especially:

  - ❑ if there is a lot of dust

  - ❑ for overhead work.

Hazards that can affect your eyes and vision 3.4

 More than 75% of eye injuries are due to something scratching or becoming embedded in the eye.

- do also wear a clear, plastic face shield when you are:
  - ❑ working with chemicals or metals that can splash
  - ❑ grinding, chipping, or using a wire brush on welds
  - ❑ undertaking work where it is likely that there will be flying particles in the air
  - ❑ sandblasting (the respirator needed for sandblasting has a helmet with a strong shield).

 Do get your eyes tested regularly (every two years) – see Section 5 for details on how to contact an optician.

- do protect your health by improving working methods with your supervisor
- do minimise damage to your health by seeking help for problems as soon as you are aware of any
- do tell your supervisors (or the site health and safety officer) **immediately** if eye protection is unsuitable, not working properly or it is not in full working order. **This is your responsibility.**

3.4 Hazards that can affect your eyes and vision

What to do if you have an eye problem

- if a dangerous or irritating chemical gets in your eye:

  - ❑ immediately start to rinse it out for at least 15 to 20 minutes straight away with flowing tap water

  - ❑ if you wear contact lenses, remove them while you are rinsing

  - ❑ get to a doctor or nurse as soon as you can

- if something hits you in your eye:

  - ❑ hold a cold compress over it, but do not press on your eye. (You can put ice cubes in a plastic bag or clean cloth.) The cold helps keep down pain and swelling

  - ❑ if pain continues or your vision is blurry, get to a doctor immediately

- if you get hit in the eye with flying metal, wood, or material from a power tool (like a drill or wheel), OR if your eye is cut or punctured:

  - ❑ **do not** wash out the eye

  - ❑ **do not** push on the eye

  - ❑ **do not** try to pull out anything that is stuck in the eye

  - ❑ get to a doctor right away

- do not wait for increased damage to be certain of your symptoms before going to your GP – talk to your supervisor, visit your doctor and explain about the work you do.

INTRODUCTION

Construction is a noisy business. Noise at high levels results in damage to the inner part of your ears, which reduces your hearing ability. Such damage is often irreversible.

The louder the noise and the longer it lasts, the more likely it is that damage will occur. Once your hearing is damaged, it usually cannot be fixed.

Damage to your ears can result in two things:

1   Loss of hearing.

2   Loss of discrimination and contrast.

So even when people shout at you, you will not be able to hear them.

One of the major problems with hearing loss is that it gradually occurs over time, so that you may not be aware of it happening. This means that it is even more important to protect yourself when you are **first exposed to noise.**

When you are at risk of damaging your hearing

- you know that the noise in the environment is at a danger level **if you need to shout to be heard by another person who is standing two metres away from you** (so you should be wearing ear protection)

3.5 Hazards that can affect your ears and hearing

- tasks which are particularly noisy, when you may be at particular risk of damage to your ears and hearing include:
  - ❑ using drills and breakers
  - ❑ operating plant machinery
  - ❑ working near compressors and generators
  - ❑ working in plant rooms, eg lift room, boiler room
  - ❑ welding
  - ❑ using woodwork machinery
  - ❑ tightening bolts with impact devices
  - ❑ driving piles

- you are at risk of damaging your ears and hearing when you are exposed to either:
  - ❑ constant background noise in the work environment (especially of long periods of time), eg machinery
  - ❑ sudden loud noises (often of very short duration), eg an explosion on a demolition site or short term use of powered tools.

 Have you been trained (or had toolbox talks) on how to **use hearing protection properly?** Ask your supervisor or health and safety rep if you need information.

Signs that tell you your hearing is being damaged

- you notice that after being in a noisy environment you have ringing in your ears, and your hearing ability is reduced. This may be a temporary or permanent problem

- you begin to have difficulty hearing sounds, especially other people talking. In speech, the letters b, k, and t become particularly hard to hear

- you may find it difficult to work out from which direction noises are coming

- gradually, the hearing loss will get worse, and you may be unable to make out what people are saying at all

- you may also have a constant ringing in your ears – this is called tinnitus

- you may have disturbed sleep and feel under stress.

 There is no cure for noise-induced hearing loss.

 Managers:

Do you comply with the requirements of the Control of Noise at Work Regulations 2005?

Do you:

1 Assess the risks from noise regularly and particularly when the work and equipment changes?

2 Assess the risk to employees subjected to noise and provide adequate hearing protection where necessary?

3 Consider effects of noise upon safe work practises (as audible warning signals and other sounds may not be heard)?

4 Seek to use quieter work methods and equipment?

5 Designate, sign and control entry to noisy areas?

6 Carry out health surveillance for workers who are at risk?

7 Comply with the other detailed requirements of the Control of Noise at Work Regulations 2005?

8 Consult with employees to ensure PPE is effective, comfortable and warm?

3.5 Hazards that can affect your ears and hearing

What you can do to prevent damage to your hearing

DO:

- wherever possible, swap jobs with other people (who are also trained in the particular task). This ensures that no one is exposed to noise for long periods

  ❑ ask your supervisor to arrange the work rota accordingly

- find out if you can use damping to reduce the noise emitted from the source.

 All hearing protection should carry the CE marking, which means that it meets the minimum safety requirements.

- use the noise control equipment that is provided in your workplace, eg hearing protection, including ear plugs and ear muffs.

- **hearing protection should be:**

  ❑ worn correctly according to the instructions

  ❑ the correct size for you

  ❑ kept clean

  ❑ inspected regularly

  ❑ replaced if damaged

  ❑ worn at ALL times.

- wear hearing protection at all times when working in noisy conditions, eg if you have difficulty hearing what a colleague is saying who is stood next to you:

  ❑ in very noisy environments, eg when using power tools, taking off your hearing protection even for a very short period can contribute to hearing problems

- protect your health by improving working methods with your supervisor

- minimise damage to your health by seeking help for problems as soon as they occur

DO NOT:

- continue to work, but tell your supervisors (or the site health and safety officer) immediately if hearing protection is not working properly or if it is not in full working order. This is **your responsibility**

- wait for increased damage to be certain of your symptoms before going to your GP – talk to your supervisor, visit your doctor and explain about the work you do.

Types of hearing protection

When either a plug or muff is properly fitted, the sound of your own voice should change, becoming  deeper, hollow, or muffled. If you don't hear the change, or if it isn't the same in both ears, you haven't obtained a proper fit in either one or both ears.

Pardon me?

Can you repeat that?

What did you say?

Use hearing protection.

3.5 Hazards that can affect your ears and hearing

How noise affects your health

"At home I was always accused of having the TV or radio on too loud. A couple of times I missed the telephone ringing. I found that there were noises I could no longer tolerate, like my wife's hair dryer. It made a **high pitched noise** and I would have to leave the bedroom if she was using it. If I were to sit quietly and read the newspaper or a book, the gas fire in the lounge, which gives off only a quiet **hissing noise**, would appear to get louder and louder until it sounded like a **jet engine** going at full throttle. I cannot even stand the humming noise a computer makes because to me the hissing noise would appear to get louder until it reaches an **intolerable** level. The problem I am experiencing with my hearing has **turned my life upside down**."

This ex-site operative is now retired. His problem was caused through workplace exposure to piling over a long period. When he first started working in construction he didn't bother wearing hearing protection. He is now unable to hear normal sounds that other people take for granted.

How the working environment can affect your health 3.6

INTRODUCTION

The environment in which you work can have a big impact on your health. This section covers:

- the temperature of your working environment
  - ❑ working in hot conditions
  - ❑ working in cold conditions
  - ❑ working in wet or damp conditions
- the air pressure of your working environment
  - ❑ the effects of working in confined spaces
- the welfare facilities of your working environment.

The working environment

Whether you are working on a small or large site, the environment can have an impact on your health, eg the weather, the cleanliness of the site.

3.6 How the working environment can affect your health

THE TEMPERATURE OF YOUR WORKING ENVIRONMENT

When working in construction, there are many times when your body may be exposed to heat or cold. This can put your body under strain.

For example:

- in high temperatures your body temperature and your heart rate may rise to uncomfortably high levels, making you feel ill

- in cold temperatures, your muscles and joints can become cold. This makes them more likely to be damaged when put under stress, such as vibration.

Working in hot conditions

When working in hot conditions the body can be at risk of heat stress. This means that the body overheats and can result in the body's core temperature and the temperature of the important internal organs, increasing to a dangerous level.

When you are at risk of heat stress

- when you are working in an environment that has little ventilation and is so hot and humid that you cannot:

  ❑ work in the environment for a long period without becoming extremely uncomfortable

  ❑ undertake your work properly

- when you are working in direct sunlight for long periods.

 Heat stress can occur in colder temperatures if the work is so physical that it makes you very hot (this is particularly true if you are wearing heavy clothing).

Heat stress may be affecting your health if:

- you feel thirsty, weak and tired and have a dry mouth

- you suffer from a headache and your skin becomes clammy

- you experience feelings of giddiness, nausea, sweating and you vomit

- you experience painful muscle cramps after physical work

- you suffer from a red, itchy heat rash.

In the worst case, you can suffer with delirium, aggressiveness, convulsions, unconsciousness, and **death**.

Working in the sun

Even when it's hot you should make sure you are wearing suitable PPE, eg high visibility vests. Without PPE you might be cooler but you are much more likely to get hit by a vehicle as you are not as visible to other workers on site. You are also protected from more of the sun's burning UV radiation.

3.6 How the working environment can affect your health

What you can do to prevent heat stress affecting your health

DO:

- drink lots of water when you get hot, even if you don't feel thirsty – your employer should provide a source of drinking water

- cool down by sponging yourself with cold water

- take regular rest breaks – if possible in a cool environment

- have a good rest at night in a cool environment (7–8 hours each night is needed by the average person).

 Death from heat stress is far more common than death from cold stress.

- wear appropriate clothing for the task and the conditions

  ❑ wear thick PPE when dealing with conductive heat, eg metals

  ❑ wear white or pale clothing when dealing with radiant heat, eg sunshine

  ❑ wear minimal amounts of clothing when working in humid conditions, eg lightweight PPE

- protect your skin from the sun – even if it's a cloudy day, you are still exposing yourself. This is because sunburned skin is already damaged and chemicals can easily get through the top layer to the more sensitive underlying tissue. You can protect your skin by:

  ❑ covering it up, eg wearing a hat or long sleeved top

  ❑ using sun cream with a high sun protection factor (SPF) (see Section 3.2)

- avoid caffeine and alcohol as these will dehydrate you further

- avoid eating large meals when working in hot conditions – this could make you feel ill as it puts extra strain on your body.

How heat stress affects your health

Asphalting is a hot job that makes you sweat. In such hot tasks, if you don't drink enough water to re-hydrate yourself, you can become unwell and develop health problems such as kidney stones. There are also other health issues shown in this picture: the operative should be wearing gloves to protect from skin complaints, eg burns, as well as wearing a hard hat and using knee protection.

- find out if it is possible to use air conditioning systems, fans or dehumidifiers to cool down your working environment

- protect your health by improving working methods with your supervisor

- minimise damage to your health by seeking help for problems as soon as you are aware of any problems

- tell your supervisors (or the site health and safety officer) **immediately** if you have any of the signs and symptoms mentioned above. This is **your responsibility**.

3.6 How the working environment can affect your health

DO NOT:

- work for long periods in hot environments
- work in hot environments without drinking water regularly
- work in the sun without protecting your skin
- ignore the symptoms
- suffer in silence.

Managers:

Check operatives' PPE and clothing are suitable and are not adding to their heat stress:

- are operatives allowed adequate breaks?
- are shaded areas provided?
- have you arranged a supply of drinking water?

Working in cold conditions

There may be many times when you may be working outside in cold, often damp, conditions.

When working in cold conditions the body can be at risk of cold stress, which means that the body gets too cold. This can result in the body's core temperature, and the temperature of the important internal organs, decreasing to a dangerous level (hypothermia).

When you are at risk of cold stress

- when you are working in an environment that has little heating and is so cold that you cannot:
  - ❑ work in the environment for a long period without becoming extremely uncomfortable
  - ❑ undertake your work properly
- when you are working in the cold for long periods.

Stress may be affecting your health

- if your skin becomes pale and white or blue in colour
- if you are unable to undertake your work as quickly as usual due to cold limbs and a lack of feeling
- if you experience painful muscle cramps after physical work
- if you have very cold fingers and toes which become numb
- if you need to urinate more frequently than normal
- if you suffer from shivering
- if, eventually, your feet and hands become frozen, indicating frostbite

In the worst case, you can suffer with disorientation, hallucinations, sleepiness, aggressiveness, unconsciousness, and death.

3.6 How the working environment can affect your health

What you can do to prevent cold stress affecting your health

DO:

- prevent heat loss by wearing a cap or hat underneath your safety helmet

- take measures to avoid the chill factor, eg wear extra clothing

- take regular rest breaks to shelter from the cold when you need to. If possible these should be taken in a warm environment

- wear appropriate clothing for the task and the conditions:
  - ❑ wear thick, warm PPE
  - ❑ pay attention to wearing adequate footwear and gloves

- be alert to areas of whiteness on the face especially the tip of the nose

- get inside and get warm if you show signs of confusion

- have a good rest at night in a warm environment

- find out if it is possible to use heating systems, eg fans, to warm up your working environment

 The construction industry is one of the top 10 industries where workers are at high risk of cold injuries.

- protect your health by improving working methods with your supervisor

- minimise damage to your health by seeking help as soon as you are aware of any problems

- tell your supervisors (or the site health and safety officer) **immediately** if you have any of the signs and symptoms mentioned above. This is **your responsibility.**

DO NOT:

- work without adequate protective clothing

- work for long periods of time in cold environments

- ignore the symptoms

- suffer in silence.

THE AIR PRESSURE OF THE WORKING ENVIRONMENT

When working in construction, there are occasions when you may be exposed to higher levels of air pressure than usual. This can put your body under stress, and cause a health problem known as decompression illness.

You are at risk of decompression illness

- when working in a tunnel and breathing compressed air of between 0.15 bar and 3.5 bar above normal pressure

- during air range diving (inland, offshore, inshore)

- during mixed gas diving (mostly offshore).

 You should be working with compressed air only if you are a specialist contractor and have the proper knowledge and training.

3.6 How the working environment can affect your health

Decompression illness may be affecting your health

- if your skin becomes itchy or mottled in appearance

- if you experience pain in your joints, especially in the knees and shoulders

- if you suffer from numbness, tingling, weakness or paralysis

- if you suffer from visual problems

- if you feel exceptionally tired and generally unwell

- if you are unable to control your bladder and bowel movements

- if, in the worst case, you suffer disorientation, convulsions and unconsciousness.

What you can do to prevent decompression illness from affecting your health

 You should not be working in areas of abnormal air pressure unless you are specifically **trained and qualified** to do so.

DO:

- ensure that your medical or certificate of fitness is up-to-date

- report to your health and safety advisor or supervisor before starting work if you:
  - ❑ are suffering from seasickness
  - ❑ are dehydrated, eg have a hangover
  - ❑ are obese.

- work out the time for which you are working at increased pressure and check your limits against the guidelines

- position air intakes so that they draw clean fresh air

- keep air compressors well maintained

- check the quality of the compressed air supply often to ensure that it is suitable

- protect your health by improving working methods with your supervisor

- minimise damage to your health by seeking help as soon as you are aware of any problems

- tell your supervisors or the site health and safety staff **immediately** if you have any of the signs and symptoms mentioned above. This is **your responsibility.**

Managers:

Plan carefully and use established decompression tables to work out depth/time or pressure/time periods.

If you are working in an area of decompression, it is law that your employer gives you regular health checks and education.

3.6 How the working environment can affect your health

DO NOT:

- work in environments of increased pressure unless you are qualified to do so

- work in increased pressure environments when suffering from any ill-health
  - ❏ this is especially dangerous if you are suffering from problems which affect your nose or your lungs.

- position compressed air intakes near to exhausts

- ignore the symptoms.

THE WELFARE FACILITIES OF YOUR WORKING ENVIRONMENT

Welfare facilities on site should include at least:

- toilets

- washing facilities and skin protection creams

- changing and rest areas

- drinking water

- eating facilities

- drying rooms.

 The law says that a person in control of a construction site must ensure that suitable and sufficient welfare facilities are provided.

Good health and welfare facilities on site will reduce health risks **but only if you use them** and help to **keep them in good condition**.

For example:

- the transmission of diseases from unwashed hands can be reduced by regular hand washing

- the incidence of dermatitis can be reduced by cleaning your skin after using harmful materials

- changing facilities can be used in cold, wet weather to change into dry PPE, reducing chances of you catching a chill.

 Wash your hands

Washing your hands is the best way to stop germs from spreading.

Think about all the things that you touched today. Colds are commonly spread by dirty hands and so are hepatitis A, meningitis, and infectious diarrhoea.

What you can do to look after your welfare

DO:

- use the facilities that are available

- clean and protect (moisturise) your hands regularly with the products provided

- change any of your wet clothes for dry ones as soon as possible

- report any of the facilities that are not working properly

- regularly clean your work clothes

- regularly clean your PPE. Ask your supervisor if your company can do this for you, especially if you are working in very dirty conditions.

3.6 How the working environment can affect your health

Hand care dispensers

Many sites now offer hand care dispenser systems that contain products to protect, clean and moisturise your hands.

Stress is the reaction people have to excessive pressure or other types of demand placed on them. Pressure in itself is not necessarily bad and many people thrive on it. It is when pressure is experienced as excessive by an individual that ill health can result.

Stress can result from:

- personal factors
- social factors
- domestic factors, or
- work factors.

Often these factors are interacting.

What is work-related stress?

Work-related stress is the reaction people have to excessive pressure or other types of demand placed on them at work.

Work related stress is not an illness, but it can lead to increased problems with ill-health, if it lasts for a long time or is particularly extreme.

Often, work related stress may occur against a background of other personal, social or domestic stress and it can be the straw that breaks the camel's back.

When you are at risk of suffering from stress

You can suffer from stress at any time. However, you are more likely to feel stressed at times when you're under pressure eg from management, or colleagues, and if you are working away from home.

3.7 How stress at work can affect your health

Signs that stress is affecting you

Stress can be noticeable in two ways:

- psychological effects
    - ❑ anxiety, eg feelings of helplessness
    - ❑ depression
    - ❑ erratic behaviour
    - ❑ emotional outbursts
- physical effects
    - ❑ heart disease, eg stroke
    - ❑ back pain
    - ❑ gastrointestinal (eg gut) problems
    - ❑ many other minor health problems, eg eczema
    - ❑ breathlessness, eg panic attacks
    - ❑ high blood pressure
    - ❑ loss of appetite
    - ❑ poor quality of sleep.

You may suffer with one or more of these symptoms.

 If you feel stressed, you're not alone. One in every five people in the UK population say they are suffering from stress.

What you can do to help reduce stress at work

At work:

- identify what the problems are and how they can be solved

- talk to your employers – if they don't know there's a problem, they can't help

- if you feel you can't talk to your employer/manager, ask a trade union (TU) representative or staff health and safety representative to raise the issue on your behalf

- you may be able to ask for an occupational health appointment

- support your colleagues if they are suffering from work-related stress and encourage them to talk to their employer/TU rep

- find out if your employer's counselling or employee assistance service (if available) can help

- speak to your GP if you are worried about your health.

(Section 2 tells you how to register with your GP and how to seek medical help).

 The law says that you must tell your employer if there are any weaknesses in their health and safety arrangements that are contributing to your stress.

Outside work:

You can help to reduce stress when you are away from work by following the guidance below:

- eat healthily

- stop smoking – it might relax you but it's bad for your overall health

- cut down on alcohol – it acts as a depressant and will not help

- cut down on caffeine (eg tea, coffee, some cold drinks) – it can make you more nervous

- avoid illegal drugs – they often make you feel even worse

- exercise – mild to moderate exercise programs keep you physically active and help you to cope with stress – exercise can motivate you and increase your energy levels

- try using some relaxation techniques, eg meditation – some people find this helps them unwind

- talk to other people about how you feel, eg your friends and family. They may be able to give you support.

It's obviously totally up to you how you live your life outside work.

However, it's worth remembering that many activities that you do outside work can affect your health. So these activities have consequences for your health while you are at work.

- see the section on **Common hazards to your health and well-being.**

You may also have other health issues to deal with that affect your overall health and capability to work.

- see the section on **Fitness for work.**

COMMON HAZARDS TO YOUR HEALTH AND WELL-BEING

Smoking

Smoking reduces your physical fitness in terms of both performance and endurance – even in fitter people. Smoking can reduce the level of maximum lung function and makes you likely to cough with phlegm or blood, suffer from shortness of breath, wheezing and gasping.

 Smoking kills around six times more people in the UK than road traffic accidents, other accidents, poisoning and overdose, murder and manslaughter, suicide and HIV infection **all put together**.

 About half of all regular cigarette smokers will eventually be killed by their habit.

3.8 How life outside work can affect your health

Consuming large amounts of alcohol

- consuming significant amounts of alcohol can lead to social or interpersonal problems. Alcohol affects many organ systems of the body, but most seriously affected are the central nervous system and the liver. Excessive alcohol use can lead to acute and chronic liver disease. The safe limits for alcohol currently recommended are 14 units (female) and 21 units (male) per week. (One unit is roughly half a pint of beer, one glass of wine or one shot of spirits.) You should ideally spread these units out over the week, having at least two alcohol-free days. Alcohol also dehydrates you so it's good to drink plenty of water after you've been drinking

 Alcohol is a major cause of liver disease (cirrhosis). After developing this condition, most people will die within five years.

- if you have been drinking a lot of alcohol during an evening, remember that you may still be over the legal limit for driving the following morning when you go to work. This can be a serious safety issue, both for you and other people around you, especially if you are operating machinery or driving a vehicle. This is, of course, also true if you drink alcohol in the morning before you go to work.

 You can still be caught for drink driving the day after a drinking session.

Listening to loud music

- going clubbing or being in other environments where there is loud music being played (eg pubs, bars or at gigs) can damage your hearing more than some types of noise at work (if hearing protection is not worn). You might think it's over the top, but wearing small, discreet ear plugs when you're out on the town can make a real difference to the protection of your hearing. Check out the table below for some facts about recommended exposure times to different noise levels.

Taking drugs (both prescribed and illegal)

- illegal drug abuse can contribute to many problems including poor health, violence, crime, and poverty. The use of prescribed drugs can sometimes also affect you badly. For example, side effects can result in dizziness, nausea, and problems with your vision, which can all affect your health and safety at work. You should check with your doctor or pharmacist if you feel that any prescribed drugs are not suiting you.

 Prescribed drugs can be dangerous at work if they make you drowsy, especially if you are using machinery. Read guidance materal carefully and follow the advice. If you have any concerns, speak to any pharmacist or your doctor.

3.8 How life outside work can affect your health

Irregular sleeping patterns

- a lack of sleep or a poor quality of sleep can reduce your ability to recall information and to think clearly. You are also less able to think about things properly, which can be significant when you have to make important decisions at work or at home. Make sure you get a proper rest as many nights as possible, especially when you are working, as this can affect your health.

 Lack of sleep can be life threatening as it reduces your levels of concentration and making you feel angry and depressed. The average amount of sleep needed by an adult each day is 7–8 hours.

Irregular eating patterns or poor eating habits

- eating properly depends on eating the right variety of food in the right amount. Too much of even the most "healthy" foods can lead to illness and disorders. It is important to eat enough to restore the energy that your body has used up (eg doing physical labour). However, eating too much or too much of the wrong types of food can result in obesity. This can put a strain on your health.

 Eating a healthy, well-balanced diet can help you feel better and live longer.

 The government recommends that we eat at least five portions of a variety of fruit and vegetables every day. A portion can be one piece of fruit, one glass of fruit juice or one serving of vegetables.

Injuries from the home and DIY

- it's amazing how many accidents happen in people's own homes. Many individuals over-estimate their ability to perform some tasks, and are determined not to be beaten by a job. In truth they need more knowledge, the right tools and someone to help! Such injuries can stop you from going to work for months.

 More than a third of all accidents to adults take place in the home. This is the largest single cause of accidents in the UK.

Injuries from sport

- injuries from sport can be caused easily and can result in some serious physical injuries. You may have to take time off from work if you've damaged yourself badly. As well as being sore and uncomfortable, your injuries may result in longer-term problems that affect your ability to do your job and your general health.

 In 1999 there were more than 740 000 sporting accidents. Football accidents accounted for 366 000 people attending hospital and 72 000 people had an accident playing rugby that resulted in a trip to A&E.

3.8 How life outside work can affect your health

Blood borne diseases

eg HIV, hepatitis B

- common risk factors for such diseases are by having sex with an infected person, receiving an infected blood transfusion or a non-sterile injection (ie sharing a needle with an infected person).

 HIV and other blood-borne diseases **cannot** be spread through skin contact. They are spread by the exchange of body fluids including blood, semen, vaginal and cervical secretions, and breast milk. So ordinary social contact such as kissing, shaking hands, coughing and sharing cutlery does not result in a blood-borne disease being passed from one person to another.

 The Human Immunodeficiency Virus (HIV) causes AIDS. There is no cure for AIDS. Although periods of illness may be interspersed with periods of remission, AIDS is almost always fatal.

Sexually transmitted infections (STIs)

eg gonorrhoea, herpes

- these infections are spread by sexual contact. Some STIs are lifelong infections, but many can be cured. STIs can cause cancers, hepatitis, cirrhosis, and other complications. Many STIs are present without symptoms. Using condoms can drastically reduce your risk of infection.

 Most STIs can be easily diagnosed and treated at Genito-Urinary Medicine (GUM) clinics which are usually based in local hospitals. Look in the phone book under GUM or sexual health.

FITNESS FOR WORK

Other factors that can also have an impact on your work include overall physical fitness and any emotional or psychological stress that you might have to cope with. This is particularly true if you are doing lots of travelling for your job. If this is the case, try to get some good quality sleep.

Some people have more long-term conditions to deal with which can have an impact on their health and safety at work.

Managers:

- do you have a system in place to assess operatives' fitness to work, in high risk jobs?

Conditions that can cause collapse or loss of consciousness

Some conditions can cause you to suddenly feel faint or collapse:

- epilepsy

- heart conditions, eg arrhythmias

- diabetes (particularly insulin dependent diabetes).

If you suffer from one of these conditions, and you have it under control, you are capable of doing most jobs on a construction site.

There is no legal requirement to tell your employer that you have one of these conditions. However, if a sudden collapse at work could cause a risk to yourself or others, you have a legal duty to report this to your supervisor or the health and safety team on your site.

3.8 How life outside work can affect your health

However, there can be issues for your safety and the safety of other people working near you if your job involves:

- working in confined spaces
- working at height, eg on scaffolding
- operating heavy plant machinery
- driving vehicles.

If you do suffer with one of these conditions, you may need special clearance from your GP to confirm your fitness to work. You may also wish to discuss your job with your supervisor or health and safety officer.

- diabetes is a condition in which the amount of glucose (sugar) in the blood is too high because the body cannot use it properly. Glucose comes from the digestion of starchy foods such as bread, rice and potatoes. The main symptoms of untreated diabetes are increased thirst, going to the toilet all the time, extreme tiredness, weight loss, and blurred vision

- epilepsy is the second most common neurologial condition after migraine, and many people suffer from some degree of epilepsy without knowing about it. Epilepsy is controlled with the correct medical treatment. However, some epileptics are prone to recurrent seizures. A seizure is caused by a temporary change in the way the brain cells work. (The old name for a seizure was a "fit"). Most seizures strike completely out of the blue. However some people have certain factors which affect them. These include alcohol, stress, patterns of light, lack of sleep, illness and food.

 Around 75% of people with epilepsy in the UK have their seizures well controlled by medication.

Pregnancy

- approximately 10 per cent of the workers on a construction site are female. So there may be times when pregnancy is an issue for health. Due to the physiological changes that the body is going through at this time, precautions should be taken regarding various types of work

  - ❏ heavy physical work such as lifting, climbing and carrying are best avoided, especially in the last three months.

 You cannot be made redundant, dismissed, or be treated unfairly because of your pregnancy.

- recent research suggests a link between job stress and problems such as miscarriage, premature delivery, low birth weight, and high blood pressure. So, if you are pregnant, it is suggested that:

  - ❏ you should work no more than eight hours a day or 40 hours a week

  - ❏ you should not stand for long periods

  - ❏ you should not lift objects heavier than 12 kilos

Other tips include:

  - ❏ taking frequent breaks

  - ❏ frequent position changes

  - ❏ putting your feet up several times a day.

- your employer should carry out an assessment of the risks and adjust your hours or conditions if needed. You should be offered a suitable alternative job if your current work can't be made safe, or else be suspended on full pay.

3.8 How life outside work can affect your health

Health conditions that can cause sensitivity to materials used at work

Construction is a hazardous business and you can be at risk from common materials used at work. Some people have existing health complaints that put them at higher risk of health problems at work.

- if you have ever suffered from **asthma**, you may be more prone to develop breathing problems (see Section 3.3)

- if you have ever suffered from eczema, you may be more prone to developing skin problems (see Section 3.2).

You can also speak to your GP and explain what your job entails.

 You should speak to your supervisor or your health and safety representative if you have any existing health complaint. They can advise you on any special measures you may need to take to keep yourself in good health.

 Be aware that you may be at increased risk of ill-health if you have any existing health complaints.

Overview

This section gives more detailed information on hazardous materials that might be used on a construction site. This is followed by a detailed bibliography that highlights the health-related rules and regulations. Finally, there is a comprehensive list of contacts that can be called upon to provide further information on heath and work related health issues.

Contents

4.1 Activities involving materials hazardous to health

This section includes details of some typical activities involving hazardous materials and should be read in association with Section 3.

 This section uses some of the chemical names of materials which you may find useful when looking at instructions or guidance notes that come with products.

The items are listed in alphabetical order of the material or operation and include the following:

- bitumen and bitumen products
- concrete work and work with mortars
- demolition
- epoxy resins
- formwork release oils and releasing agents
- facade cleaning
- flooring work
- glass cleaning
- isocyanates
- oils and lubricants
- paint removal, pickling
- paint work
- stone conditioners and stone impregnating agents
- synthetic mineral fibres
- tiling work
- timber (Parquet) flooring
- timber preservation.

Bitumen and bitumen products

Products based on bitumen are used for roadside paving and construction purposes whenever sealing or water-tight components are required. Main products include:

- bituminous roof covering and sealing materials

- bituminous paints (emulsions and solutions)

- sealing compounds

- poured asphalt floor topping and others.

Today, bitumen is no longer classified as potentially carcinogenic since it is no longer blended with tar products. The main aspects of the health hazard of bitumen are the hydrocarbon mixes emitted in the form of steam or aerosol during the handling of the hot product, particularly in closed locations.

With bituminous paint or sealants, the contact of bitumen with the skin could lead to acne-like skin diseases and dermatitis (see Section 3.2).

 These materials will have warning notices and instructions – use them!

Concrete work and work with mortars

A strongly alkaline mixture forms when adding water to cement and/or lime, bringing the risk of caustic burns. In addition, the cement in concrete has a sensitising effect due to its alkalinity and chromium content. See Section 3.2 for information on how to avoid problems.

4.1 Activities involving materials hazardous to health

Demolition and excavation

The range of materials which could be met during demolition and excavation (particularly in land-fill situations) is immense and great care must be taken. Expert investigation and assessment is required and expert back-up and support must be available during the work.

Epoxy resins

Increasingly epoxy resin products are used in the construction industry, mainly due to their technical characteristics, namely chemical resistance and mechanical strength. The main uses are coatings, eg flooring in industrial structures, reworking of concrete structures, adhesives for tiles and flooring materials. Problems in handling epoxy resins are mainly skin allergies. Contact allergies are found mostly on hands and lower arms. The amine-containing hardeners often have an irritating, caustic effect. Apart from this irritation, the resin components may lead to skin sensitisation (see Section 3.2).

 If in doubt, stop work and seek expert advice.

Formwork release oils and releasing agents

Releasing agents (formwork release oils and formwork releasing agents) are special auxiliary materials used to reliably separate the formwork from partly or completely cured concrete. A film on the formwork surface prevents adhesion between the formwork and the concrete.

The type of application may be important for the respective health hazard:

- when applied manually with a cloth, sponge, spatula, brush or roller the risk of contact with the skin is particularly high, and this method should be avoided

- when applied by broom the distance to the formwork surface is sufficiently wide. The danger of splashing is particularly high when applying the product above the head.

Working overhead should also be avoided when applying the products with a spraying device as this may additionally involve the danger of hazardous aerosols. The state-of-the-art technology is application by a fully-mechanical device which performs all work sequences by simply changing the respective pads (ie cleaning and application of the release agent). With this technology, potential contact with the release agent is less likely than with other methods.

 The use of proper PPE is essential to reduce these risks.

The following health risks must be considered:

- product splashing into the eyes may cause conjunctivitis. A variety of skin troubles can be caused by contact with the release oils (irritations of the skin, formation of eczemas, removal of the skin's natural fat)

- inhaling vapours and mists may lead to irritation of the respiratory tract and damage your health. In enclosed locations, hazardous accumulations of gases and vapours may occur. Inhalation should be avoided at all cost, since chronic damage to the respiratory organs cannot be excluded.

4.1 Activities involving materials hazardous to health

Façade cleaning

Façade cleaning includes the cleaning of anodised or paint-coated metal façades, natural or synthetic stone, facing plaster, plastic and wood facings, etc. Primarily, this work covers the removal of dust, soot, slight corrosion as well as oil and grease but also includes paint and graffiti.

The cleaners used can be divided mainly into strongly acid or alkaline cleaners as well as solvent-laden cleaners. The health hazard of products with a very high or very low pH lies mainly in their caustic characteristics. When solvent-laden cleaners are used, a number of protective measures have to be observed. Often, these products contain dichloromethane – a substance suspected to be potentially carcinogenic.

Flooring work

Fine grained quartz dust from the grinding of the floor topping may cause silicosis. Chromate-containing concrete-type stoppers may result in skin diseases, while solvent-laden primers and adhesives may cause headaches, tiredness, numbness and even unconsciousness.

Glass cleaning

Some glass cleaners contain caustic and toxic hydrofluoric acid. They are used to clean strongly contaminated glass fronts (corrosion). A number of accident prevention measures must be observed when handling these products.

Isocyanates

Many products used in the construction industry contain polyurethane plastics, eg insulating materials, coatings, paints, adhesives, wood floor sealants, erection foams, joint fillers and sealing compounds.

Isocyanates are primarily known for their irritating effect on the eyes, skin (contact allergies) and mucous membranes. In addition, they are sensitizing, ie they may cause allergies upon inhaling or skin contact – particularly "isocyanate asthma".

 The symptoms of these respiratory tract problems such as coughing and shortness of breath may lead to more severe health problems.

Oils and lubricants

Fats, oils, hydraulic oils and lubricants are required for the maintenance of the various construction machines and transport vehicles.

Health hazards from inhaling vapours mainly arise only when large amounts have been spilled and must be removed and disposed of.

Damage to the skin would primarily occur when the products have been spilled or inadequately handled.

 Some materials are carcinogenic, so avoid skin contact, including oily rags in overall pockets.

4.1 Activities involving materials hazardous to health

Paint removal, pickling

Strongly alkaline pickling solutions may cause caustic burns of the skin and pickling solutions containing chlorinated hydrocarbons may irritate the skin, eyes and mucous membranes as well as damage the central nervous system. Moreover, chlorinated hydrocarbons are carcinogenic (they give you cancer).

Removal of old paints containing lead is likely to involve the release of fine particles containing lead which may be ingested or breathed in. Work methods must be designed to avoid this and PPE used rigorously. Residues must be cleared away and disposed of in a controlled manner. Note that children are particularly susceptible to damage from lead.

Paint work

Depending on the type of organic binder used (alkyd resin, acrylic resin, synthetic resin), the paints and coatings contain solvents and/or water, pigments, fillers and other additives. The main hazards in the paint and varnishing trade lie in the solvent-dilutable paints. The following aromatic hydrocarbons may be found, among others: ethyl benzene, xylene, isopropyl benzene and mesitylene. The total content of aromatic solvents ranges between one and 10 per cent. Toluene and benzene may be found as contaminants.

 The health hazards are found when inhaling solvent vapours or aerosols, by ingestion, intake through the skin and direct contact with the eyes.

Stone conditioners and stone impregnating agents

Stone compactors are products to compact/condition, weathered and crumbling stone façades. Apart from solvents (acetone, butanone, ethanol and toluene or hydrocarbon mixtures) they also contain an organic silicic acid ester. Furthermore, they may also contain organic tin compounds (toxic, skin-resorptive) and alkylalkoxysiloxanes. In addition to the hazardous effects of the solvent the tetraethylsilicate may cause damage to the lungs and larynx.

Stone impregnators are products that provide mineral substrates with a water, oil, colour and contaminant-proofing coat.

The health hazard derives mainly from the high solvent concentration.

Synthetic mineral fibres

 Synthetic mineral fibres mainly cover two groups:

- mineral wool insulation material (glass wool, mineral wool, 98% market share)
- ceramic fibre products (2% market share).

Both groups contain breathable fibres which will irritate and are potentially carcinogenic. (However, the effect of fibre-type particles in the lungs has not yet been fully researched.)

Tiling work

Even floor tilers are affected by the use of hazardous materials. The main impact is the damage to the skin from handling cement-containing products and reaction resins (may cause allergies). However, irritation of the respiratory tract caused by solvent-laden fillers (acetic acid, 2-methylethyl ketone oxime, methanol, 2-methoxyethanol, butylamines, cyclohexylamine and others) can be found too. 2-methoxyethanol constitutes a serious health hazard because it affects the genes, ie it can impair fertility and affect the health of foetuses.

Timber (parquet) flooring

Dust from oak and beech wood, from the grinding of parquet floor and other timber flooring may lead to the formation of malignant tumours inside the nose (adenocarcinomes). Other wood dust types are suspected to be carcinogenic. Solvent-laden wood mastics and agents for surface treatment may irritate the respiratory and digestion tracts, cause vertigo, intoxication, headaches and numbness.

4.1 Activities involving materials hazardous to health

Timber preservation

Water-soluble wood preservatives may contain inorganic boron compounds, silicofluorides (skin irritation), hydrogen fluorides (caustic burns), chromium (VI) compounds (carcinogenic in the form of breathable dusts/aerosols) and hazardous copper compounds.

Oily wood preservatives may contain solvent mixes, tar oils and derivatives of coaltar (skin irritation and skin cancer).

Useful contacts 4.2

This section gives the contact details of some useful organisations that can provide you with information about health at work.

Local health authorities

- health authorities are responsible for determining a wide range of health-related services in your local community

- one of their most important functions is to help you find a GP and other medical practitioners

- to find a local GP visit website: <http://www.nhs.uk/servicedirectories/Pages/ServicesSearch.aspx>

First aid societies

- St John Ambulance
  27 St John's Lane
  London EC1M 4BU
  02073 244000
  <http://www.sja.org.uk/sja>

- St. Andrews Ambulance Association
  National Headquarters
  St. Andrew's House
  48 Milton Street
  Glasgow G4 0HR
  01413 324031
  <http://www.firstaid.org.uk>

4.2 Useful contacts

Health & Safety Executive

- for all your enquiries on workplace health and safety:

  HSE InfoLine
  Telephone 08453 450055
  <http://www.hse.gov.uk/contact>

  HSE Information Services
  Caerphilly Business Park
  Caerphilly CF83 3GG

- to report an accident at work:

  RIDDOR reporting
  Telephone 0845 300 9923
  <http://www.hse.gov.uk/riddor>
  Incident Contact Centre

  Caerphilly Business Park
  Caerphilly CF83 3GG

- to order HSE publications:

  HSE Books
  PO Box 1999
  Sudbury
  Suffolk CO10 2WA
  Telephone 01787 881165
  <http://www.hsebooks.com/books>

Local law centres

- law centres provide a free and independent professional legal service to people who live or work in their catchment areas

- law centres were to set up to overcome the obstacles faced by people who need access to the legal system. Free, publicly provided legal advice should be available to everyone, not just to those with financial resources or to those few that can get legal aid because of their income. Also there are many areas of law where legal aid is simply not available. This means that even in areas where fundamental rights are in dispute there is no access to the legal system. Legal aid is not available, for example, for representation at industrial tribunals or immigration appeals tribunals

- law centres specialise in those areas of law including welfare rights, immigration and nationality, housing and homelessness, employment rights, and sex and race discrimination

- law centres are all around the UK. Contact the Law Centres Federation to find your nearest branch:

The Law Centres Federation
Duchess House
18–19 Warren Street
London W1T 5LR
02074 284400
Contact through the website at:
<http://www.lawcentres.org.uk>

4.2 Useful contacts

Citizens Advice Bureau

- the Citizens Advice Bureau Service offers free, confidential, impartial and independent advice

- every Citizens Advice Bureau (CAB) is a registered charity reliant on volunteers. Citizens Advice Bureaux help solve nearly six million new problems every year which are central to people's lives, including debt and consumer issues, benefits, housing, legal matters, employment, and immigration. Advisers can help fill out forms, write letters, negotiate with creditors and represent clients at court or tribunal

- there are 2000 CAB outlets in England, Wales and Northern Ireland. Each CAB is an independent charity, relying on funding from the local authority and from local business, charitable trusts and individual donations

- to access advice go to: <http://www.adviceguide.org.uk>

- to find your local CAB office go to: <http://www.citizensadvice.org.uk/getadvice.htm>

CIRIA

- CIRIA is recognised as the foremost independent and authoritative broker of construction research and innovation in the UK

- to find out more contact:

  CIRIA
  Classic House
  174–180 Old Street
  London EC1V 9BP, UK
  Phone: +44 (0) 20 7549 3300
  Fax: +44 (0) 20 7253 0523
  <www.ciria.org>
  enquiries@ciria.org

Construction Industry Training Board (CITB)

- CITB has a complete range of services available to help you be safer including advice, publications and courses. Whether you are an employer, a health and safety adviser or a site supervisor, CITB offers a wide range of advice tailored to your individual needs

- CITB produces a range of quality health and safety publications, from printed reference documents to videos and computer-based programs. Companies and individuals can tap into a wide range of expertise through the National Construction College, which offers a wide variety of courses to suit your needs. You can also find out all you need to know about the CSCS card schemes through: <http://www.citb.co.uk/cardschemes/whatcardschemesareavailable/cscs/>

- to find out more contact:

  CITB Construction Skills Head Office
  Bircham Newton
  Kings Lynn
  Norfolk PE31 6RH
  Phone: 01485 577577. Fax: 01485 577793
  Email: information.centre@citb.co.uk
  Visit: <http://www.citb.co.uk/>

4.2 Useful contacts

The Trade and General Workers Union

- with more than 800 000 members in every type of workplace, the T&G is the UK's biggest general union with a long and proud tradition of representing members in the workplace

- thanks partly to union campaigning, Britain has comprehensive health and safety legislation. The T&G provides updated safety, health and environmental information in its monthly T&G Record magazine

- there are a variety of publications available from the trade unions about looking after your health, statutory sick pay, disablement benefit, and claiming damages

- further information is available from:

T&G Central Office
Transport House
128 Theobalds Road
Holborn
London WC1X 8TN
Phone: 020 7611 2500
<http://www.tgwu.org.uk>

Construction Confederation (CC)

- the CC is the leading representative body for contractors, representing some 5000 companies who, in turn, are responsible for over 75 per cent of the industry's turnover

- the CC are able to offer members quick and practical advice about issues or problems they encounter surrounding a wide range of issues including health and safety

- to find out more contact:

The Construction Confederation
55 Tufton Street
Westminster
London SWIP 3QL
08708 989090
enquiries@thecc.org.uk

4.3 References

Legislation

The Health and Safety at Work etc Act 1974

Health and Safety (First Aid) Regulations 1981

Manual Handling Operations Regulations 1992

Personal Protective Equipment at Work Regulations 1992

The Workplace (Health, Safety and Welfare) Regulations 1992

Reporting of Injuries, Diseases and dangerous Occurrences Regulations (RIDDOR) 1995

The Health And Safety (Consultation With Employees) Regulations (HSCER) 1996

The Confined Spaces Regulations 1997

The Provision and Use of Work Equipment Regulations 1998

The Lifting Operations and Lifting Equipment Regulations 1998

Management of Health and Safety at Work Regulations1999

The Ionising Radiation Regulations 1999

Control of Substances Hazardous to Health Regulations (COSHH) 2002

The Control of Lead at Work Regulations 2002

The Control of Noise at Work Regulations 2005

The Control of Vibration at Work Regulations 2005

Chemicals (Hazard Information and Packaging for Supply) Regulations (CHIP or CHIPS) 2002 as amended 2005

The Control of Asbestos Regulations 2006

Construction (Design and Management) Regulations 2007

Approved Codes of Practice

L5	Control of substances hazardous to health
L21	Management of health and safety at work
L22	Safe use of work equipment
L24	The Workplace (Health, Safety and Welfare) Regulation 1992
L25	The Personal Protection at Work Regulations
L95	Health and Safety – Consultation with Employees
L96	A guide to the Work in Compressed Air Regulations
L132	The Control of Lead at Work Regulations
L101	Safe work in confined spaces
L108	Controlling Noise at Work
L112	Safe use of power presses
L113	The Lifting Operations and Lifting Equipment Regulations
L114	Safe use of woodworking machinery
L121	The Ionising Radiation Regulations
L140	Hand-arm vibration
L141	Whole-body vibration
L143	The Control of Asbestos Regulations
L144	Managing health and safety in construction

4.3 References

GUIDANCE

HSE guidance notes

EH40   Workplace exposure limits

EH43   Carbon monoxide

EH46   Man-made mineral fibres

EM series   Asbestos essentials

MS24   Medical aspects of occupational skin disease

MS25   Medical aspects of occupational asthma S46 *In situ* timber
            treatment using timber

MSA8   Arsenic and you

 The HSE guidance notes, booklets, leaflets and construction information sheets are available from the HSE website and many can also be downloaded

HSE booklets

HSG53	Respiratory Protective Equipment (RPE) – a practical guide for users (2nd edition 1991)
HSG54	The maintenance, examination and testing of local exhaust ventilation
HSG61	Surveillance of people exposed to health risks at work
HSG88	Hand-arm vibration
HSG97	A step by step guide to COSHH assessments
HSG137	Health risk management
HSG150	Health and safety in construction
HSG167	Biological monitoring in the workplace
HSG173	Monitoring strategies for toxic substances
HSG193	COSHH Essentials: easy steps to control chemicals
HSG202	General ventilation in the workplace
HSG210	Asbestos essentials
HSG247	Asbestos: the Licenced Contrators Guide
HSG248	Asbestos: the analysts' guide for sampling, analysis and clearance procedures
HSG251	Control of substances hazardous to health in fumigation operations

4.3 References

HSE leaflets

INDG84	Leptospirosis – are you at risk
INDG90	Understanding Ergonomics at work
INDG91	Drug misuse at work
INDG95	Respiratory sensitisers and COSHH: breathe freely
INDG147	Keep your top on: health risks from working in the sun
INDG171	Upper limb disorders in the workplace: risk factor checklist
INDG175	Control the risk from hand-arm vibration: advice for employers on the Control of Vibration at Work Regulations 2005
INDG197	Working with sewage: The health hazards
INDG233	Preventing contact dermatitis at work
INDG240	Don't mix it: a guide for employers on alcohol at work
INDG281	Working together to reduce stress at work
INDG362	Noise at work
INDG383	Manual handling assessment charts

HSE construction information sheets

8   –   *Safety in excavations*

10   –   *Tower scaffolds*

18   –   *Provision of welfare facilities at fixed construction sites*

26   –   *Cement*

27   –   *Solvents*

36   –   *Silica health hazards*

45   –   *Fire safety*

46   –   *Welfare provisions on transient construction sites*

47   –   *Inspection and reports*

54   –   *Dust control on concrete cutting saws*

57   –   *Handling kerbs*

Other references

Bielby, S C and Gilbertson, A L (2008)
Site safety handbook (fourth edition)
CIRIA C669, London (ISBN 978 0 86017 669 5)

Steeds, J E, Shepherds, E and Barry, D L (1996)
A guide for safe working on contaminated sites
CIRIA R132, London (ISBN 978 0 86017 451 6)

For your notes